D0948885

London Mathematical Society
Lecture Note Series 39

Affine Sets and Affine Groups

D. G. NORTHCOTT

CAMBRIDGE UNIVERSITY PRESS

LONDON MATHEMATICAL SOCIETY LECTURE NOTE SERIES

Already published in this series

1. General cohomology theory and K-theory, PETER HILTON.
4. Algebraic topology: A student's guide, J. F. ADAMS.
5. Commutative algebra, J. T. KNIGHT.
7. Introduction to combinatory logic, J. R. HINDLEY, B. LERCHER, and J. P. SELDIN.
8. Integration and harmonic analysis on compact groups, R. E. EDWARDS.
9. Elliptic functions and elliptic curves, PATRICK DU VAL.
10. Numerical ranges II, F. F. BONSALL and J. DUNCAN.
11. New developments in topology, G. SEGAL (ed.).
12. Symposium on complex analysis Canterbury, 1973, J. CLUNIE and W. K. HAYMAN (eds.).
13. Combinatorics, Proceedings of the British combinatorial conference 1973, T. P. McDONOUGH and V. C. MAVRON (eds.).
14. Analytic theory of abelian varieties, H. P. F. SWINNERTON-DYER.
15. An introduction to topological groups, P. J. HIGGINS.
16. Topics in finite groups, TERENCE M. GAGEN.
17. Differentiable germs and catastrophes, THEODOR BRÖCKER and L. LANDER.
18. A geometric approach to homology theory, S. BUONCRISTIANO, C. P. ROURKE and B. J. SANDERSON.
19. Graph theory, coding theory and block designs, P. J. CAMERON and J. H. VAN LINT.
20. Sheaf theory, B. R. TENNISON.
21. Automatic continuity of linear operators, ALLAN M. SINCLAIR.
22. Presentations of groups, D. L. JOHNSON.
23. Parallelisms of complete designs, PETER J. CAMERON.
24. The topology of Stiefel manifolds, I. M. JAMES.
25. Lie groups and compact groups, J. F. PRICE.
26. Transformation groups: Proceedings of the conference in the University of Newcastle upon Tyne, August 1976, CZES KOSNIOWSKI.
27. Skew field constructions, P. M. COHN.
28. Brownian motion, Hardy spaces and bounded mean oscillation, K. E. PETERSEN.
29. Pontryagin duality and the structure of locally compact abelian groups, SIDNEY A. MORRIS.
30. Interaction models, N. L. BIGGS.
31. Continuous crossed products and type III von Neumann algebras, A. VAN DAELE.
32. Uniform algebras and Jensen measures, T. W. GAMELIN.
33. Permutation groups and combinatorial structures, N. L. BIGGS and A. T. WHITE.
34. Representation theory of Lie groups, M. F. ATIYAH.
35. Trace ideals and their applications, BARRY SIMON.

36. Homological group theory, edited by C. T. C. WALL.
37. Partially ordered rings and semi-algebraic geometry, GREGORY W. BRUMFIEL.
38. Surveys in combinatorics, edited by B. BOLLOBAS.

London Mathematical Society Lecture Note Series. 39

Affine Sets and Affine Groups

D. G. Northcott

Professor of Pure Mathematics

University of Sheffield

CAMBRIDGE UNIVERSITY PRESS

CAMBRIDGE

LONDON NEW YORK NEW ROCHELLE

MELBOURNE SYDNEY

Published by the Press Syndicate of the University of Cambridge
The Pitt Building, Trumpington Street, Cambridge CB2 1RP
32 East 57th Street, New York, N.Y. 10022, USA
296 Beaconsfield Parade, Middle Park, Melbourne 3206, Australia

First published 1980

Printed in Great Britain by
Redwood Burn Ltd., Trowbridge and Esher

Library of Congress Cataloguing in Publication Data
Northcott, Douglas Geoffrey
Affine sets and affine groups
(London Mathematical Society lecture note series; 39
ISSN 0076-0552)
1. Geometry, Affine. 2. Set theory.
I. Title. II. Series.
516'.4 QA477 79-41595

ISBN 0 521 22909 X

Contents

page

Introduction ix

PART I. AFFINE SETS

Chapter 1 Preliminaries concerning algebras 3
- 1.1 Algebras 3
- 1.2 Subalgebras and factor algebras 4
- 1.3 Examples of K-algebras 5
- 1.4 Tensor products of vector spaces 6
- 1.5 Tensor products of algebras 7
- 1.6 Enlargement of the ground field 9

Chapter 2 Affine sets 12
- 2.1 Rational maximal ideals 12
- 2.2 Function algebras 17
- 2.3 Loci and the associated topology 19
- 2.4 Affine sets 24
- 2.5 Examples of affine sets 28
- 2.6 Morphisms of affine sets 33
- 2.7 Products of function algebras 37
- 2.8 Products of affine sets 40
- 2.9 Some standard morphisms 42
- 2.10 Enlargement of the ground field 46
- 2.11 Generalized points and generic points 55

Chapter 3 Irreducible affine sets 60
- 3.1 Irreducible spaces 60
- 3.2 Irreducible affine sets 63
- 3.3 Localization 70
- 3.4 Rational functions and dimension 74
- 3.5 Enlargement of the ground field 79
- 3.6 Almost surjective K-morphisms 84

Chapter 4 Derivations and tangent spaces 95
- 4.1 Derivations in algebras 95
- 4.2 Examples of derivations 101
- 4.3 Derivations in fields 104

4.4 Tangent spaces and simple points 114
4.5 Tangent spaces and products 124
4.6 Differentials 125

PART II. AFFINE GROUPS

Chapter 5 Affine groups 133
5.1 Affine groups 133
5.2 Components of an affine group 143
5.3 Examples of affine groups 148
5.4 K-homomorphisms of affine groups 159
5.5 K-morphic actions on an affine set 162
5.6 G-modules 166
5.7 General rational G-modules 173
5.8 Linearly reductive affine groups 175
5.9 Characters and semi-invariants 181
5.10 Linearly reductive affine groups and invariant theory 189
5.11 Quotients with respect to linearly reductive groups 196
5.12 Quotients with respect to finite groups 202
5.13 Quotient groups of affine groups 206

Chapter 6 The associated Lie algebra 222
6.1 General K-algebras 222
6.2 Lie algebras 224
6.3 The Lie algebra of an affine group 227
6.4 Extension of the ground field 241
6.5 A basic example 243
6.6 Further examples 250
6.7 Adjoint representations 255
Chapter 7 Power series and exponentials 263
7.1 Rings of formal power series 263
7.2 Modules over a power series ring 267
7.3 Exponentials 271
7.4 Applications to affine groups 279

References 283

Index 284

Introduction

The topics treated in the following pages were largely covered in two seminars, both given at Sheffield University, one during the session 1976/7 and the other during 1978/9. I had noted sometime earlier that M. Hochster and J. A. Eagon had established a connection between Determinantal Ideals and Invariant Theory. However in order to understand what was involved I had first to acquaint myself with the relevant parts of the theory of Algebraic Groups. With this in mind, I began to read J. Fogarty's book on Invariant Theory.

Almost at once my interest broadened. It had been my experience to see Commutative Algebra develop out of attempts to provide classical Algebraic Geometry with proper foundations, but it had been a matter of regret that the algebraic machinery created for this purpose tended to conceal the origins of the subject. Fogarty's book helped me to see how one could look at Geometry from a readily accessible modern standpoint that was still not too far removed from the kind of Coordinate Geometry which now belongs to the classical period of the subject.

When my own ideas had reached a sufficiently advanced stage I decided to try and develop them further by committing myself to giving a seminar. With a subject such as this, and in circumstances where the time available was very limited, it was necessary to assume a certain amount of background knowledge. Indeed most accounts of aspects of the theory of Algebraic Groups assume a very great deal in the way of prerequisites. In my case the audience could be assumed to be knowledgeable about Commutative Noetherian rings and I planned the lectures with this in mind. The outcome is that the following treatment is very nearly self-contained if one presupposes a knowledge of field theory, tensor products and the more familiar parts of Commutative Algebra including, of course, the famous Basis Theorem and Zeros Theorem of Hilbert. It is true that there are a few places where additional background knowledge is required, but the reader who knows the topics mentioned above will find that other results are rarely used, and that where they are it will usually require little time and effort to fill in the gaps. To help him I have provided suitable references wherever they are likely to be needed.

The book falls naturally into two parts with Chapters 1-4 forming the first part. Here the aim is to show how those loci in classical analytical geometry that are defined by the solutions of simultaneous algebraic equations (together with the appropriate transformations of one such object into another) can be turned into a category. In this context the ambient affine space which makes geometrical thinking possible has to be removed from the picture, but not so far that it cannot be brought back readily when geometrical insight into a situation is needed.

Turning now to points of detail, Chapter 1 is used to explain certain matters that have to do with terminology and notation. It is also used to give a brief survey of the properties of tensor products over a field. The latter enables the discussion of (i) products of affine algebraic sets, and (ii) the consequences of enlarging the ground field, to take place later without an interruption to explain technicalities of a purely algebraic natu

The development of geometrical ideas begins with Chapter 2. Besi describing affine sets, their products and their morphisms, I have also revived the theory of specializations because, it seems to me, it provide techniques that are both interesting and highly effective. Chapter 3 introduces the concept of the irreducible components of an affine set, the idea of dimension, and the very important topic of almost surjective morphisms. The last of the chapters devoted to Geometry deals with the subject of tangent spaces and simple points.

Part 1 contains all the Geometry that is needed for the reader to be able to understand the rest of the book. But although it was planned with the idea of supporting an introduction to Algebraic Groups it is the author hope that it will be of interest to those whose main concern is to get an insight into the foundations of Algebraic Geometry. If the first four chapters are read with this more limited end in view, then I would sugge ending with section (4.5) because the last section of the chapter consists of technical material used later to study the connection between an Affine

Group and its associated Lie Algebra. A natural continuation of the geometrical sections would be the theory of tangent bundles.

The second part moves fairly quickly into the study of Affine Groups. These are groups which also have an affine structure and where the group operations are compatible with this structure. After the definitions and a discussion of the relation between connectedness and irreducibility, such topics as rational representations, linearly reductive groups, and the beginnings of invariant theory are considered. Chapter 5 ends with a comparatively elementary proof that every factor group of an affine group, with respect to a closed normal subgroup, has itself a natural structure as an affine group. This is one place where the account draws more heavily on the reader's knowledge of Commutative Algebra than it does elsewhere, but the topic seemed of sufficient importance to justify a departure from the guide lines which I had set myself for the book as a whole. It may help the reader to know that Chapter 5 makes very little use of Chapter 4. If therefore he wishes to get to the second part as quickly as possible, he may prefer to begin Chapter 5 after completing Chapter 3, and then to return to Chapter 4 to fill in certain gaps at a later stage. This in fact is what I did in my seminars.

The theory of Lie Algebras is not introduced until Chapter 6. Here it is shown that with each Affine Group there is associated a Lie Algebra and a detailed study is made of some of the most important examples. It is possible to exploit this connection very effectively in the case of a ground field of characteristic zero. The final chapter provides the theory which in this case enables properties of the associated Lie Algebras to be transferred back to the Affine Groups. It is here that the account stops. To continue further it would be necessary to include a short course on Lie Algebras after which an account of classical affine groups could be given.

I have, of course, made use of the writings of other mathematicians. I have already mentioned the special debt I owe to J. Fogarty's book on Invariant Theory. I also made considerable use of C. Chevalley's first account of Algebraic Groups, and the presentation, by H. Bass, of a course given by A. Borel on Linear Algebraic Groups. In the case of Geometry, it was a pleasure to re-read part of A. Weil's book on the

Foundations of Algebraic Geometry. So far as Commutative Algebra is concerned I have relied principally on the books of O. Zariski, P. Samuel M. Nagata, and my own writings. Detailed references will be found in the text and in the booklist given at the end.

On a more personal level I would like to thank those who came to my seminars and encouraged me by the interest they showed, and once again it is a pleasure to thank Mrs. E. Benson for doing all the typing with such excellent results.

Finally I would like to express my appreciation to the London Mathematical Society and the Syndics of the Cambridge University Press for agreeing to let this appear in their series of Lecture Notes.

Sheffield
February 1979

D. G. Northcott

Part I. Affine Sets

1 · Preliminaries concerning algebras

General remarks

In order to describe the structure of affine algebraic sets we need to use the theory of algebras, so this initial chapter will be devoted to a survey of some of the basic properties of algebras that will be needed shortly. However for the present it will suffice to restrict our attention to those algebras which are associative and have identity elements. These will be referred to as unitary, associative algebras.

There is one convention we shall use to which it is necessary to draw attention. Normally any ring we consider will have an identity element. When dealing with such rings the term ring-homomorphism is always understood to mean a mapping which is compatible with addition and multiplication and which also takes identity element into identity element. In other terms our ring-homomorphisms are assumed to be unitary. Throughout Chapter 1 K will be used to denote a field. It is not assumed that K is algebraically closed nor even that it contains infinitely many elements.

1.1 Algebras

Let R be a ring which is also a vector space over the field K. R is said to be an associative K-algebra if

$$(ka)b = a(kb) = k(ab)$$

for all $a, b \in R$ and $k \in K$. If in addition R possesses an identity element 1_R, then R is described as a unitary, associative K-algebra.

Suppose that R is a unitary, associative K-algebra. Then the mapping

$$\omega : K \to R$$

defined by $\omega(k) = k1_R$ is a (unitary) ring-homomorphism which maps K into the centre of R. This ring-homomorphism is referred to as the *structural homomorphism* of the algebra. If R is *non-trivial*, that is if its identity element is not zero, this allows us to identify K with a subring of the centre of R.

Let us reverse the procedure. Suppose that R is a ring with identity element and that we are given a ring-homomorphism $\omega : K \to R$ which maps K into the centre of R. If for $a \in R$ and $k \in K$ we put $ka = \omega(k)a$, then R becomes a unitary, associative K-algebra having ω as its structural homomorphism. Both ways of looking at these algebras will be useful.

Let R and R' be unitary, associative K-algebras and $\phi : R \to R'$ a (unitary) ring-homomorphism. We say that ϕ is a *homomorphism of K-algebras* if it is also a K-linear mapping of R into R'. Thus if $\omega : K \to R$ and $\omega' : K \to R'$ are the structural homomorphisms, then ϕ is a homomorphism of K-algebras if and only if $\phi \circ \omega = \omega'$. In the case where R and R' are both non-trivial we may regard K as being embedded in each of the two rings. In this situation ϕ is a homomorphism of K-algebras when and only when it leaves the elements of K fixed.

An isomorphism of K-algebras is, of course, a homomorphism of K-algebras which is also a bijection. Every isomorphism has an inverse which is also an isomorphism.

1.2 Subalgebras and factor algebras

Let R be a unitary, associative K-algebra and let S be a subring of R. (It is understood that the identity element of R is shared with all its subrings.) If S is also a K-subspace of R, then S has the structure of a unitary, associative K-algebra and the inclusion mapping $j : S \to R$ is a homomorphism of K-algebras. We describe this situation by saying that S is a *subalgebra* of the K-algebra R. Evidently the intersection of any family of subalgebras of R is again a subalgebra.

Let A be a set of elements of R. There will be a smallest subalgebra S (say) which contains A. We say that S *is generated by* A. The typical element of S is a linear combination (with coefficients in K) of finite products $a_1 a_2 \dots a_n$ $(n \geq 0)$, where $a_i \in A$. Naturally the empty product is understood to have the value 1_R. If $S = R$, then A is said to be a system of generators for R. If R can be generated by a finite set, then it is said to be finitely generated as a K-algebra.

Now let $\mathfrak{A}$ be a two-sided ideal of R. Then $\mathfrak{A}$ is a K-subspace of R and the ring $R/\mathfrak{A}$ has a unique structure as a unitary, associative K-algebra subject to the requirement that the natural mapping of R on to $R/\mathfrak{A}$ shall be a homomorphism of K-algebras. Whenever we refer to $R/\mathfrak{A}$ as a K-algebra it is always this structure that we have in mind.

1.3 Examples of K-algebras

All the examples of K-algebras given in section (1.3) are unitary and associative.

Example 1. K itself is a K-algebra the structural homomorphism in this case being the identity mapping. It follows that if R is any unitary, associative K-algebra, then the structural homomorphism $\omega : K \to R$ is a homomorphism of K-algebras.

Example 2. Any ring R with identity element which contains K as a subring of its centre is, of course, a K-algebra. In particular any extension field of K is a K-algebra.

Example 3. Let V be a vector space over K. (The dimension of V need not be finite.) The endomorphisms of K form a K-algebra which we denote by $\mathrm{End}_K(V)$. Here the structural homomorphism is obtained by mapping the element k, of K, into the endomorphism that consists in multiplication by k.

Example 4. Let n be a positive integer and denote by $M_n(K)$ the set of all $n \times n$ matrices with entries in K. These form a K-algebra. To obtain the structural homomorphism we map the element of k, of K, into the corresponding scalar matrix.

Let V be an n-dimensional vector space over K. Then $M_n(K)$ and $\mathrm{End}_K(V)$ are isomorphic K-algebras.

Example 5. Let $X_1, X_2, \ldots, X_n$ be indeterminates. Then the polynomial ring $K[X_1, X_2, \ldots, X_n]$ is a commutative K-algebra.

Example 6. Let V be a non-empty set and denote by $\mathfrak{F}_K(V)$ the set of all mappings of V into K. If f, g belong to $\mathfrak{F}_K(V)$ we define their <u>sum</u> and <u>product</u> by

$$(f + g)(v) = f(v) + g(v)$$

and

$$(fg)(v) = f(v)g(v).$$

This turns $\mathfrak{F}_K(V)$ into a commutative K-algebra where the structural homomorphism identifies $k \in K$ with the corresponding constant function.

1.4 Tensor products of vector spaces

Before we examine the important concept of the tensor product of two K-algebras we shall say a few words about tensor products of vector spaces.

Throughout section (1.4) U, V and W will denote vector spaces over K. (Their dimensions need not be finite.) All tensor products will be taken over K so we shall use $\otimes$ instead of the more ornate $\otimes_K$.

As is well known $U \otimes V$ is a vector space over K its typical element being a finite sum of tensor products $u \otimes v$, where $u \in U$ and $v \in V$. A fundamental property of $U \otimes V$ is the following: <u>given a bilinear mapping</u>

$$\omega : U \times V \to W,$$

<u>there is a unique</u> K-<u>linear mapping</u> $\overline{\omega} : U \otimes V \to W$ <u>such that</u> $\overline{\omega}(u \otimes v) = \omega(u, v)$ <u>for all</u> u <u>in</u> U <u>and</u> v <u>in</u> V.

Let $\{u_i\}_{i \in I}$ be a base for the space U and $\{v_j\}_{j \in J}$ a base for V. Then the elements $u_i \otimes v_j$ $(i \in I,\ j \in J)$ form a base for $U \otimes V$. Thus if A is a linearly independent subset of U and B a linearly independent subset of V, then the elements $a \otimes b$ $(a \in A,\ b \in B)$ form a linearly independent subset of $U \otimes V$. In particular if $u \in U$, $u \neq 0$ and $v \in V$, $v \neq 0$, then $u \otimes v \neq 0$.

The following result is often useful.

Theorem 1. _Let_ $u_1, u_2, \ldots, u_n$ _belong to_ U _and_ $v_1, v_2, \ldots, v_n$ _to_ V. _Suppose that_ $u_1, u_2, \ldots, u_n$ _are linearly independent over_ K _and that_

$$u_1 \otimes v_1 + u_2 \otimes v_2 + \ldots + u_n \otimes v_n = 0.$$

Then $v_i = 0$ _for_ $i = 1, 2, \ldots, n$.

Remark. Of course the roles of U and V may be interchanged.

Proof. The theorem becomes obvious as soon as one expresses $v_1, v_2, \ldots, v_n$ in terms of a base for V over K.

Now let $\phi : U \to U'$ and $\psi : V \to V'$ be homomorphisms of vector spaces. These give rise to a homomorphism

$$\phi \otimes \psi : U \otimes V \to U' \otimes V'$$

in which $u \otimes v$ is mapped into $\phi(u) \otimes \psi(v)$. Obviously if ϕ and ψ are surjections, then $\phi \otimes \psi$ is a surjection. But K is a _field._ Consequently if ϕ and ψ are injections, then $\phi \otimes \psi$ is also an injection.

Let A be a subspace of U and B a subspace of V. The inclusion mappings $A \to U$ and $B \to V$ induce an _injection_ $A \otimes B \to U \otimes V$. This enables us to regard $A \otimes B$ as a subspace of $U \otimes V$. Next the natural homomorphisms $U \to U/A$ and $V \to V/B$ give rise to a surjective homomorphism

$$U \otimes V \to (U/A) \otimes (V/B)$$

and it is readily checked that

$$\operatorname{Ker}\{U \otimes V \to (U/A) \otimes (V/B)\} = (A \otimes V) + (U \otimes B). \tag{1.4.1}$$

1.5 Tensor products of algebras

In this section the letters R and S will be used to denote unitary, associative K-algebras. As in the last section we shall use $\otimes$ to stand for a tensor product taken over K.

If we form $R \otimes S$, then this is not simply a K-space. It has in fact the structure of a unitary, associative K-algebra in which multiplication satisfies

$$(r_1 \otimes s_1)(r_2 \otimes s_2) = r_1 r_2 \otimes s_1 s_2, \tag{1.5.1}$$

where the notation is self-explanatory. The identity element of this algebra is $1_R \otimes 1_S$. Furthermore $R \otimes S$ and $S \otimes R$ are isomorphic K-algebras under an isomorphism which matches $r \otimes s$ with $s \otimes r$.

Consider the mapping $R \to R \otimes S$ in which $r \mapsto r \otimes 1_S$. This is a homomorphism of K-algebras and in the case where S is non-trivial (and therefore $1_S \neq 0_S$) it is an injection. Accordingly *when S is non-trivial we may regard R as a subalgebra of the K-algebra* $R \otimes S$. Naturally there is a corresponding homomorphism $S \to R \otimes S$ to which similar observations apply.

K itself plays a special role in the theory of tensor products. This manifests itself through the fact that we always have an isomorphism of K-algebras

$$R \otimes K \approx R \tag{1.5.2}$$

in which $r \otimes k$ is matched with kr.

Now assume that $\phi : R \to R'$ and $\psi : S \to S'$ are homomorphisms of K-algebras. Then

$$\phi \otimes \psi : R \otimes S \to R' \otimes S'$$

is a homomorphism of K-algebras. Let $\mathfrak{A}$ be a two-sided ideal of R, $\mathfrak{B}$ a two-sided ideal of S, and let us apply the above observation to the natural mappings $R \to R/\mathfrak{A}$ and $S \to S/\mathfrak{B}$. Since these are surjective homomorphisms of K-algebras, they give rise to

$$R \otimes S \to (R/\mathfrak{A}) \otimes (S/\mathfrak{B}) \tag{1.5.3}$$

which is itself a surjective homomorphism of K-algebras. Put

$$[\mathfrak{A}, \mathfrak{B}] = \mathfrak{A} \otimes S + R \otimes \mathfrak{B} \tag{1.5.4}$$

this being regarded in the first instance as a subspace of $R \otimes S$. By (1.4.1) it is none other than the kernel of (1.5.3). It follows therefore that $[\mathfrak{A}, \mathfrak{B}]$ is a two-sided ideal of $R \otimes S$ and moreover we have an isomorphism

$$(R \otimes S)/[\mathfrak{A}, \mathfrak{B}] \approx (R/\mathfrak{A}) \otimes (S/\mathfrak{B}) \tag{1.5.5}$$

of K-algebras.

Lemma 1. *Let the notation be as above and suppose that* $\mathfrak{A} \neq R$, $\mathfrak{B} \neq S$. *Then the inverse images of* $[\mathfrak{A}, \mathfrak{B}]$ *with respect to the canonical homomorphisms* $R \rightarrow R \otimes S$ *and* $S \rightarrow R \otimes S$ *are* $\mathfrak{A}$ *and* $\mathfrak{B}$ *respectively.*

This is clear because, as previously observed, $[\mathfrak{A}, \mathfrak{B}]$ is the kernel of (1.5.3).

Finally we observe that if $\{a_i\}_{i \in I}$ is a system of generators for the K-algebra R and $\{b_j\}_{j \in J}$ is a system of generators for the K-algebra S, then the elements

$$a_i \otimes 1_S \quad \text{and} \quad 1_R \otimes b_j,$$

where $i \in I$ and $j \in J$, form a system of generators for $R \otimes S$. In particular *if* R *and* S *are finitely generated* K-*algebras so too is* $R \otimes S$.

1.6 Enlargement of the ground field

Throughout section (1.6) we shall assume that L is an *extension field* of K.

First suppose that V is a vector space over K and put

$$V^L = V \otimes_K L. \tag{1.6.1}$$

Then V^L has a natural structure as a vector space over L in which

$$\lambda'(v \otimes \lambda) = v \otimes \lambda'\lambda \tag{1.6.2}$$

it being understood that $v \in V$ and $\lambda, \lambda' \in L$. Naturally V^L can also be regarded as a K-space.

Consider the mapping $V \to V^L$ in which $v \mapsto v \otimes 1_L$. This is K-linear and it is also an injection. Consequently we may (and usually do) regard V as a K-subspace of V^L. On this understanding any base for V over K is a base for V^L over L; in particular, if the elements $v_1, v_2, \ldots, v_n$ of V are linearly independent over K, then they are also linearly independent over L. Furthermore to a certain extent the roles of V and L can be interchanged. To be precise let $\{\lambda_i\}_{i \in I}$ be a base for L over K. If now $\omega \in V^L$, then there exists a unique family $\{w_i\}_{i \in I}$ of elements of V such that (i) only a finite number of the w_i are non zero, and (ii) $\omega = \Sigma \lambda_i w_i$.

Next let $f : U \to V$ be a homomorphism of K-spaces. This will have a unique extension to a homomorphism $U^L \to V^L$ of L-spaces. In fact the extension of f is just $f \otimes L$.

Now suppose that R and S are unitary, associative K-algebras. Then $R^L = R \otimes L$ is both a vector space over L and at the same time a ring with an identity element. Indeed R^L is a unitary, associative L-algebra in which the structural homomorphism $L \to R^L$ maps the element λ, of L, into $1_R \otimes \lambda$.

Our earlier remarks show that we can regard R as being embedded in R^L. If this is done, then R is a subalgebra of R^L when the latter is regarded as a K-algebra. Moreover any set of elements of R which generates R as a K-algebra will also generate R^L as an L-algebra. In particular, _when R is a finitely generated K-algebra, R^L is a finitely generated L-algebra._

Once again let $\{\lambda_i\}_{i \in I}$ be a base for L over K and suppose that $\beta \in R^L$. Our previous observations show that there exists a unique family $\{b_i\}_{i \in I}$ of elements of R such that $\beta = \Sigma \lambda_i b_i$. Hence if $\mathfrak{A}$ is a two-sided ideal of R and $\mathfrak{A} R^L$ is the two-sided ideal it generates in R^L, then $\beta \in \mathfrak{A} R^L$ if and only if $b_i \in \mathfrak{A}$ for all $i \in I$. It follows that

$$\mathfrak{A} R^L \cap R = \mathfrak{A}. \tag{1.6.3}$$

Also if $\{\mathfrak{A}_j\}_{j \in J}$ is a family of two-sided ideals of R, then

$$\bigcap_{j \in J} \left(\mathfrak{A}_j R^L\right) = \left(\bigcap_{j \in J} \mathfrak{A}_j\right) R^L. \qquad (1.6.4)$$

Finally assume that $\phi : R \to S$ is a homomorphism of K-algebras. Then the homomorphism $R^L \to S^L$, of L-spaces, which it induces is actually a homomorphism of L-algebras. For example, suppose that $\mathfrak{A}$ is a two-sided ideal of R. Then the natural homomorphism $R \to R/\mathfrak{A}$ induces a surjective homomorphism $R^L \to (R/\mathfrak{A})^L$ of L-algebras, that is a homomorphism $R \otimes L \to (R/\mathfrak{A}) \otimes L$. By (1.4.1) the kernel of this is $\mathfrak{A} \otimes L = \mathfrak{A} R^L$ and therefore we have an isomorphism

$$R^L / \mathfrak{A} R^L \approx (R/\mathfrak{A})^L \qquad (1.6.5)$$

of L-algebras.

Although considerably more could be said about algebras and tensor products in the general context of this chapter, we have now reached a convenient point at which to begin concentrating the discussion on the situations that are particularly relevant to the treatment of affine algebraic sets.

2 · Affine sets

General remarks

In this chapter the conventions and notation remain the same as those introduced in the introductory paragraphs to Chapter 1. In particular K will denote a completely arbitrary field and the symbol $\otimes$ will be used to denote a tensor product taken over K when there is no possibility of a misunderstanding. In other situations, for example when it is necessary to form a tensor product over a different field, a subscript will be attached to $\otimes$ in order to make the intention quite clear.

We shall now place further restrictions on the algebras we study. Whereas Chapter 1 was principally concerned with K-algebras that were unitary and associative, throughout this chapter those that will occupy our attention will usually be commutative as well.

One further point. We shall have occasion to consider polynomials in indeterminates $X_1, X_2, \ldots, X_n$ with coefficients in K. By the null polynomial will be meant the polynomial all of whose coefficients are zero. Now if $f(X_1, X_2, \ldots, X_n)$ belongs to the polynomial ring $K[X_1, X_2, \ldots, X_n]$, then it determines a function, $\bar{f}$ say, from the n-fold product $K \times K \times \ldots \times K$ to K. We shall call $\bar{f}$ a polynomial function. It is necessary to bear in mind that $\bar{f}$ may be the zero function (that is all of its values may be zero) even though $f(X_1, X_2, \ldots, X_n)$ is not the null polynomial.

2.1 Rational maximal ideals

Throughout section (2.1) R and S will denote K-algebras which are unitary, associative and commutative. Naturally the tensor product $R \otimes S$, taken over K, inherits the same three properties.

Let M be a maximal ideal of R. Then R/M is a field containing K as a subfield.

Definition. M will be said to be a 'rational maximal ideal' or (more precisely) to be 'K-rational' if $R/M = K$ or (equivalently) if R/M and K are isomorphic K-algebras.

The set of K-rational maximal ideals of R will be denoted by $\mathfrak{M}_K(R)$.

Theorem 1. _Let_ $\phi : S \to R$ _be a homomorphism of_ K-_algebras and_ M _a rational maximal ideal of_ R. _Then_ $\phi^{-1}(M)$ _is a rational maximal ideal of_ S.

Proof. We may regard $S/\phi^{-1}(M)$ as a subalgebra of the non-trivial K-algebra R/M, and then

$$K \subseteq S/\phi^{-1}(M) \subseteq R/M = K.$$

The theorem follows.

In the case where ϕ is _surjective_ the relation between the rational maximal ideals of R and those of S is very simple. We record the facts in the following

Corollary. _Let_ $\phi : S \to R$ _be a surjective homomorphism of_ K-_algebras and let_ M _be a maximal ideal of_ R. _Then_ $\phi^{-1}(M)$ _is a maximal ideal of_ S. _Furthermore if either_ M _or_ $\phi^{-1}(M)$ _is_ K-_rational, then so is the other._

Proof. We have an isomorphism $S/\phi^{-1}(M) \approx R/M$ of K-algebras and all the assertions follow from this.

Example. Let $X_1, X_2, \ldots, X_n$ be indeterminates, where $n \geq 1$, and put

$$K^n = K \times K \times \ldots \times K.$$

Here it is understood that the right hand side is an ordinary cartesian product and there are n factors. Thus a typical element of K^n is a sequence $(\alpha_1, \alpha_2, \ldots, \alpha_n)$, where $\alpha_i \in K$, and it is clear that we have a bijection between

$$K^n \text{ and } \mathfrak{M}_K(K[X_1, X_2, \ldots, X_n])$$

which matches $(\alpha_1, \alpha_2, \ldots, \alpha_n)$ with the rational maximal ideal $(X_1-\alpha_1, X_2-\alpha_2, \ldots, X_n-\alpha_n)$.

We mention, in passing, that when K is algebraically closed, every maximal ideal of $K[X_1, X_2, \ldots, X_n]$ is K-rational.† This is an important result with far-reaching consequences, but they will concern us only to a small extent in Chapter 2.

Suppose, by way of contrast, that K is a finite field with q elements. Then the non-zero elements of K form a multiplicative group of order $q - 1$ and therefore $\alpha^q = \alpha$ for all $\alpha \in K$. In particular the polynomial $X_1^q - X_1$ belongs to all the rational maximal ideals of $K[X_1, X_2, \ldots, X_n]$ and, in consequence, the intersection of these maximal ideals is not zero. However, when K is an infinite field the situation is quite different as we shall now show.

Theorem 2. Let Ω be an infinite subset of K, and let $\phi(X_1, X_2, \ldots, X_n)$ be non-null and belong to $K[X_1, X_2, \ldots, X_n]$. Then there exist $\omega_1, \omega_2, \ldots, \omega_n$ in Ω with the property that $\phi(\omega_1, \omega_2, \ldots, \omega_n) \neq 0$.

Proof. We use induction on n and begin by observing that the assertion is certainly true when $n = 1$. Now assume that $n > 1$ and that the theorem has been proved for polynomials involving only $n - 1$ variables.

We can write

$$\phi(X_1, X_2, \ldots, X_n) = \sum_{j=0}^{m} \phi_j(X_1, X_2, \ldots, X_{n-1})X_n^{m-j}, \qquad (2.1.1)$$

where $\phi_j(X_1, X_2, \ldots, X_{n-1}) \in K[X_1, X_2, \ldots, X_{n-1}]$ and $\phi_0(X_1, X_2, \ldots, X_{n-1})$ is not null. The inductive hypothesis shows that there exist $\omega_1, \omega_2, \ldots, \omega_{n-1}$ in Ω for which $\phi_0(\omega_1, \omega_2, \ldots, \omega_{n-1}) \neq 0$; and then the case of a single variable shows that we can find ω_n, in Ω,

---

† See [(9) Theorem 6, p. 285]. The number in round brackets refers to the list of references at the end.

so that $\phi(\omega_1, \ldots, \omega_{n-1}, \omega_n) = \sum_{j=0}^{m} \phi_j(\omega_1, \ldots, \omega_{n-1})\omega_n^{m-j}$ is different from zero.

Corollary. If K is an infinite field and $n \geq 1$, then the intersection of all the K-rational maximal ideals of $K[X_1, X_2, \ldots, X_n]$ is zero.

At this stage it is convenient to introduce a new definition. We shall say that the K-algebra R is rationally reduced with respect to K if

$$\bigcap_{M \in \mathfrak{M}_K(R)} M = (0).$$

This enables us to reorganize some of the information derived above.

Theorem 3. Suppose that $n \geq 1$. Then the polynomial ring $K[X_1, X_2, \ldots, X_n]$ is rationally reduced with respect to K if and only if K is infinite.

We shall now take a quick look at what happens to R and its rational maximal ideals if we enlarge the ground field from K to L (say). Certain general comments on this operation were made in section (1.6). These will be used here. Note that R^L is now a commutative L-algebra.

Theorem 4. Let M be a rational maximal ideal of the K-algebra R, and let L be an extension field of K. Then MR^L is a rational maximal ideal of the L-algebra R^L.

Proof. By (1.6.5) and (1.5.2), we have isomorphisms

$$R^L/MR^L \approx (R/M) \otimes_K L \approx K \otimes_K L \approx L$$

of L-algebras and from these the theorem follows.

Corollary 1. If R is rationally reduced with respect to K, then R^L is rationally reduced with respect to L.

Proof. If M denotes a typical rational maximal ideal of R, then the intersection of all the M's is zero. It follows, from (1.6.4), that the R^L-ideals MR^L have zero as their intersection, so a fortiori the full set of L-rational maximal ideals also has intersection zero.

From section (1.6) we know that a homomorphism $\phi : S \to R$ of K-algebras induces a homomorphism $S^L \to R^L$ of L-algebras.

Corollary 2. _Let_ $\phi : S \to R$ _be a homomorphism of_ K-_algebras and_ M _a_ K-_rational maximal ideal of_ R. _Then the inverse image of_ MR^L _with respect to the induced homomorphism_ $S^L \to R^L$ _is_ $(\phi^{-1}(M))S^L$. _This is an_ L-_rational maximal ideal of_ S^L.

Proof. By Theorems 1 and 4, $(\phi^{-1}(M))S^L$ is an L-rational maximal ideal of S^L and it is obviously mapped into MR^L. The corollary follows.

We next turn our attention to rational maximal ideals in a tensor product. To this end let M be a rational maximal ideal of R, N a rational maximal ideal of S, and define [M, N] as in (1.5.4). Then [M, N] is an ideal of $R \otimes S$ and, by (1.5.5), we have isomorphisms

$$(R \otimes S)/[M, N] \approx (R/M) \otimes (S/N) \approx K \otimes K \approx K$$

of K-algebras. It follows that [M, N] is a rational maximal ideal of $R \otimes S$. Thus there is a mapping

$$\mathfrak{M}_K(R) \times \mathfrak{M}_K(S) \to \mathfrak{M}_K(R \otimes S)$$

in which (M, N) is mapped into [M, N]. Moreover, by Chapter 1 Lemma 1, the mapping is an injection.

Now let I be a rational maximal ideal of $R \otimes S$ and denote by M' respectively N' the inverse image of I with respect to the canonical homomorphism $R \to R \otimes S$ respectively $S \to R \otimes S$. It follows (Theorem 1) that M' is a rational maximal ideal of R and N' a rational maximal ideal of S. Accordingly [M', N'] is a maximal ideal of $R \otimes S$ and since $[M', N'] \subseteq I$ we have $[M', N'] = I$. This proves

Theorem 5. _There is a bijection between_ $\mathfrak{M}_K(R) \times \mathfrak{M}_K(S)$ _and_ $\mathfrak{M}_K(R \otimes_K S)$ _which matches_ (M, N) _of the former with_ [M, N] _of the latter._

2.2 Function algebras

Let V be a set and as in section (1.3) denote by $\mathfrak{F}_K(V)$ the commutative K-algebra formed by all the mappings of V into K. Now let R be a subalgebra of the K-algebra $\mathfrak{F}_K(V)$. Thus R is a K-algebra whose elements are K-valued functions on V. To describe this situation we shall say that (V, R) is a function algebra over K. Note that such a function algebra is a unitary, associative and commutative K-algebra, and that it contains all constant K-valued functions with domain V.

For the rest of section (2.2) it will be assumed that (V, R) is a given function algebra over K. Let $x \in V$. Then there exists a surjective homomorphism $R \to K$ of K-algebras in which $f \mapsto f(x)$. Put

$$M_x = \{f \mid f \in R \text{ and } f(x) = 0\}. \tag{2.2.1}$$

Then M_x is the kernel of the homomorphism and therefore it is a maximal ideal of R. Indeed from the isomorphism $R/M_x \approx K$ we see that M_x is a rational maximal ideal. Furthermore any f which belongs to all the M_x $(x \in V)$ must vanish everywhere, and therefore it must be the zero element of the algebra. This proves

Theorem 6. *Every function algebra over K is rationally reduced with respect to K.*

We return to the consideration of (V, R). If $x \in V$ and $f \in R$, then $f - f(x)$ belongs to M_x, and therefore we have

$$f(x) = \text{the } M_x\text{-residue of } f. \tag{2.2.2}$$

Next we have a mapping $V \to \mathfrak{M}_K(R)$ given by $x \mapsto M_x$. Usually this is neither an injection nor a surjection. However we do have

Lemma 1. *Let $x, y \in V$. Then the following statements are equivalent:*

(a) $M_x \neq M_y$;

(b) *there exists* $f \in R$ *such that* $f(x) = 0$ *and* $f(y) \neq 0$;

(c) *there exists* $g \in R$ *such that* $g(y) = 0$ *and* $g(x) \neq 0$.

Proof. Assume (a). Then M_x is not contained in M_y. Consequently there exists $f \in M_x$ such that $f \notin M_y$. This shows that (a) implies (b) and similarly it implies (c). Obviously each of (b) and (c) implies (a) so the proof is complete.

Function algebras arise quite naturally. Suppose that S is some given unitary, associative, and commutative K-algebra. For each $\phi \in S$ we can define a mapping

$$\overline{\phi} : \mathfrak{M}_K(S) \to K$$

by

$$\overline{\phi}(M) = \text{the } M\text{-residue of } \phi. \tag{2.2.3}$$

The set of all $\overline{\phi}$'s forms a subalgebra $\overline{S}$ (say) of $\mathfrak{F}_K(\mathfrak{M}_K(S))$; that is to say $(\mathfrak{M}_K(S), \overline{S})$ is a function algebra over K. We shall call it the derived function algebra of S. Note that we have a surjective homomorphism $S \to \overline{S}$ of K-algebras in which $\phi \mapsto \overline{\phi}$. We shall speak of this as the canonical homomorphism of S on to its derived function algebra.

Lemma 2. The kernel of the canonical homomorphism $S \to \overline{S}$ is the intersection of all the rational maximal ideals of S. Consequently the canonical homomorphism is an isomorphism of K-algebras if and only if S is rationally reduced.

This is clear because $\overline{\phi} = 0$ if and only if $\phi \in M$ for every M in $\mathfrak{M}_K(S)$. Note that if M is a rational maximal ideal of S, then $M\overline{S}$ is a rational maximal ideal of $\overline{S}$. In fact we have the

Corollary. There is a bijection between $\mathfrak{M}_K(S)$ and $\mathfrak{M}_K(\overline{S})$ in which M in $\mathfrak{M}_K(S)$ is matched with $M\overline{S}$ in $\mathfrak{M}_K(\overline{S})$.

Note that when $(\mathfrak{M}_K(S), \overline{S})$ is regarded as a function algebra and $M \in \mathfrak{M}_K(S)$, then the corresponding rational maximal ideal of $\overline{S}$ in the sense of (2.2.1) is $M\overline{S}$. Thus, in this particular instance, we get a bijection between the domain of the function algebra and the set of its rational maximal ideals.

Example. Consider the derived function algebra of $K[X_1, X_2, \ldots, X_n]$, where $X_1, X_2, \ldots, X_n$ are indeterminates and $n \geq 1$. We have already seen, in section (2.1), that there is a natural bijection between K^n and the set of rational maximal ideals. If we now identify these, then the derived function algebra may be written as $(K^n, \overline{K[X_1, \ldots, X_n]})$.

Suppose next that $\phi(X_1, X_2, \ldots, X_n)$ belongs to $K[X_1, X_2, \ldots, X_n]$ and let $\overline{\phi}$ be its image in $\overline{K[X_1, X_2, \ldots, X_n]}$. Then $\overline{\phi} : K^n \to K$ and, by (2.2.3),

$$\overline{\phi}(\alpha_1, \alpha_2, \ldots, \alpha_n) = \phi(\alpha_1, \alpha_2, \ldots, \alpha_n).$$

Thus $\overline{K[X_1, X_2, \ldots, X_n]}$ _consists of the members of_ $K[X_1, X_2, \ldots, X_n]$ _now regarded as functions from_ K^n _to_ K. _Observe that, by Lemma 2 and Theorem 3, the canonical homomorphism_

$$K[X_1, X_2, \ldots, X_n] \to \overline{K[X_1, X_2, \ldots, X_n]}$$

is an isomorphism when and only when the field K _is infinite._

2.3 Loci and the associated topology

Throughout section (2.3) (V, R) denotes a given function algebra over K. As before we set

$$M_x = \{f \mid f \in R \text{ and } f(x) = 0\}$$

so that M_x is a K-rational maximal ideal of R. Now suppose that A is a subset of R and put

$$C_V(A) = \{x \mid x \in V \text{ and } f(x) = 0 \text{ whenever } f \in A\}. \tag{2.3.1}$$

Thus $C_V(A)$ consists of the common zeros of the functions which make up A. We call $C_V(A)$ the _locus_ of the set A. Note that if AR denotes the ideal generated by A, then

$$C_V(A) = C_V(AR). \tag{2.3.2}$$

Also if A_1 and A_2 are subsets of R, then

$$A_1 \subseteq A_2 \text{ implies } C_V(A_2) \subseteq C_V(A_1). \tag{2.3.3}$$

Suppose that $f \in R$. It will be convenient to write $C_V(f)$ rather than $C_V(\{f\})$. Thus

$$C_V(f) = \{x \mid x \in V \text{ and } f(x) = 0\}. \tag{2.3.4}$$

A locus of this kind is sometimes called a <u>principal locus.</u> Since

$$C_V(A) = \bigcap_{f \in A} C_V(f) \tag{2.3.5}$$

it follows that <u>every locus is an intersection of principal loci and vice versa.</u>

Let $\mathfrak{A}$, $\mathfrak{B}$ be ideals of R and, in addition, let $\{\mathfrak{A}_i\}_{i \in I}$ be a family of ideals of the algebra. It is easily verified that

$$\begin{aligned} &C_V(0) = V \text{ and } C_V(1) = \emptyset \\ &C_V(\mathfrak{A}\mathfrak{B}) = C_V(\mathfrak{A}) \cup C_V(\mathfrak{B}), \\ &C_V(\sum_{i \in I} \mathfrak{A}_i) = \bigcap_{i \in I} C_V(\mathfrak{A}_i). \end{aligned} \tag{2.3.6}$$

This shows that V <u>can be endowed with a topology in which the closed sets are precisely the sets</u> $C_V(\mathfrak{A})$, <u>i. e. the closed sets in the topology are the various loci.</u> We shall refer to this as the R-<u>topology</u> on V.

Let $\mathfrak{A}$ be an ideal of R. In preparation for the next set of observations we define $\mathrm{Rad}_V(\mathfrak{A})$ as the intersection of all the rational maximal ideals of the form M_x, where $x \in V$, that contain $\mathfrak{A}$. (If there are no such maximal ideals containing $\mathfrak{A}$, then it is to be understood that $\mathrm{Rad}_V(\mathfrak{A}) = R$.) Evidently

$$\mathfrak{A} \subseteq \mathrm{Rad}_V(\mathfrak{A}) \tag{2.3.7}$$

and

$$\mathrm{Rad}_V(\mathrm{Rad}_V(\mathfrak{A})) = \mathrm{Rad}_V(\mathfrak{A}). \tag{2.3.8}$$

Also if $\mathfrak{A}_1$ and $\mathfrak{A}_2$ are ideals, then

$$\mathfrak{A}_1 \subseteq \mathfrak{A}_2 \text{ implies } \operatorname{Rad}_V(\mathfrak{A}_1) \subseteq \operatorname{Rad}_V(\mathfrak{A}_2). \tag{2.3.9}$$

Lemma 3. _Let_ $\mathfrak{A}$ _be an ideal of_ R. _Then_ $\mathfrak{A} = \operatorname{Rad}_V(\mathfrak{A})$ _if and only if some subfamily of_ $\{M_x\}_{x \in V}$ _has_ $\mathfrak{A}$ _as its intersection._

This is clear from the definition of $\operatorname{Rad}_V(\mathfrak{A})$ when it is taken in conjunction with (2.3.7).

Once again let $\mathfrak{A}$ be an ideal. Then

$$C_V(\mathfrak{A}) = \{x \mid x \in V \text{ and } \mathfrak{A} \subseteq M_x\} \tag{2.3.10}$$

which shows at once that

$$C_V(\mathfrak{A}) = C_V(\operatorname{Rad}_V(\mathfrak{A})). \tag{2.3.11}$$

Theorem 7. _Let_ $\mathfrak{A}$ _and_ $\mathfrak{B}$ _be ideals of_ R. _Then_ $C_V(\mathfrak{A}) = C_V(\mathfrak{B})$ _if and only if_ $\operatorname{Rad}_V(\mathfrak{A}) = \operatorname{Rad}_V(\mathfrak{B})$.

Proof. If $C_V(\mathfrak{A}) = C_V(\mathfrak{B})$, then the relation $\operatorname{Rad}_V(\mathfrak{A}) = \operatorname{Rad}_V(\mathfrak{B})$ follows from (2.3.10). The converse is a consequence of (2.3.11). Note that (2.3.11) and Theorem 7 together show that $\operatorname{Rad}_V(\mathfrak{A})$ _is the largest ideal with the same locus as_ $\mathfrak{A}$.

We shall now consider subsets of V and seek to find a useful criterion for determining when such a subset is a locus. To this end suppose that $U \subseteq V$ and put

$$I_V(U) = \{f \mid f \in R \text{ and } f(x) = 0 \text{ whenever } x \in U\}. \tag{2.3.12}$$

Evidently $I_V(U)$ is an ideal of R. (If U is empty we put $I_V(U)$ equal to R.) It will be called the _associated ideal_ of the subset U. Now

$$I_V(U) = \bigcap_{x \in U} M_x \tag{2.3.13}$$

and therefore, by Lemma 3,

$$\operatorname{Rad}_V(I_V(U)) = I_V(U). \tag{2.3.14}$$

Again, by (2.3.13) and (2.3.10),

$$I_V(C_V(\mathfrak{A})) = \mathrm{Rad}_V(\mathfrak{A}) \tag{2.3.15}$$

whenever $\mathfrak{A}$ is an ideal of R, and from this follows

Theorem 8. *Let $\mathfrak{A}$ be an ideal of R. Then $\mathfrak{A} = \mathrm{Rad}_V(\mathfrak{A})$ if and only if $I_V(C_V(\mathfrak{A})) = \mathfrak{A}$.*

Before proceeding to the next theorem let us note for future reference that if U_1, U_2 are subsets of V, then

$$U_1 \subseteq U_2 \text{ implies } I_V(U_2) \subseteq I_V(U_1). \tag{2.3.16}$$

Theorem 9. *Let V be endowed with the R-topology, let U be a subset of V, and let $\bar{U}$ be the closure of U in V. Then*

$$C_V(I_V(U)) = \bar{U}$$

and therefore U is closed if and only if $C_V(I_V(U)) = U$. In any case $I_V(U) = I_V(\bar{U})$.

Proof. Since $C_V(I_V(U))$ is closed and contains U, we have $\bar{U} \subseteq C_V(I_V(U))$. Further, because $\bar{U}$ is closed, $\bar{U} = C_V(\mathfrak{A})$ for some ideal $\mathfrak{A}$. Now $f(\bar{U}) = \{0\}$ and hence $f(U) = \{0\}$ for every $f \in \mathfrak{A}$. Thus $\mathfrak{A} \subseteq I_V(U)$ and therefore

$$C_V(I_V(U)) \subseteq C_V(\mathfrak{A}) = \bar{U}$$

as required.

It remains for us to prove the final assertion. However $I_V(\bar{U}) = I_V(C_V(I_V(U))) = \mathrm{Rad}_V(I_V(U))$ by (2.3.15), and so $I_V(\bar{U}) = I_V(U)$ by (2.3.14).

Corollary. *Let U_1 and U_2 be subsets of V. Then $I_V(U_1) = I_V(U_2)$ if and only if $\bar{U}_1 = \bar{U}_2$.*

The next theorem summarizes some of the most important consequences of our discussion so far.

Theorem 10. Let U denote a typical closed subset of V and $\mathfrak{A}$ a typical ideal of R satisfying $\mathfrak{A} = \mathrm{Rad}_V(\mathfrak{A})$. Then there is a natural bijection connecting the set of U's and the set of $\mathfrak{A}$'s. This is such that if U is matched with $\mathfrak{A}$, then $I_V(U) = \mathfrak{A}$ and $C_V(\mathfrak{A}) = U$. Moreover the bijection reverses inclusion relations.

Example. Let X be an indeterminate. Then the derived function algebra of K[X] is $(K, \overline{K[X]})$, where the notation is the same as that employed in section (2. 2). The typical member of $\overline{K[X]}$ is a polynomial g(X) considered as a function defined on K. Accordingly the corresponding principal locus consists of the zeros of g(X) in K. This can be an arbitrary finite set or, in the case where g(X) is the null polynomial, the whole of K.

If S is any set, then there is a topology on S in which the closed subsets are S itself and all its finite subsets. Let us call this the finite sets topology on S. Since every locus is an intersection of principal loci, the remarks of the last paragraph show that the $\overline{K[X]}$-topology on K is precisely the finite sets topology.

Once again suppose that (V, R) is a function algebra over K. Let $f \in R$ and let $\{k_1, k_2, \ldots, k_s\}$ be a finite subset of K. Put $g = (f-k_1)(f-k_2) \ldots (f-k_s)$. Then $g \in R$ and the inverse image of $\{k_1, k_2, \ldots, k_s\}$ with respect to f is the principal locus $C_V(g)$. This is a closed subset of V when V has the R-topology.

Theorem 11. Let K be endowed with the finite sets topology. Then the R-topology on V is the weakest topology in which the mappings $f : V \to K$, that constitute R, are continuous.

Proof. Assume V has a topology in which every member of R is continuous. The theorem will be proved if we show that the various loci in V are closed in the given topology, and this will follow if we establish that the typical principal locus $C_V(f)$, where $f \in R$, is closed. But this is clear because $C_V(f)$ is the inverse image of the finite set whose only member is the zero element of K.

2.4 Affine sets

Let (V, R) be a function algebra over K, let $x \in V$, and as previously put $M_x = \{f | f \in R \text{ and } f(x) = 0\}$. We shall say that (V, R) is an algebraic affine set defined over K or, when there is no need to be quite so explicit, an affine set provided both the following conditions are satisfied:

(a) the mapping $V \to \mathfrak{M}_K(R)$ given by $x \mapsto M_x$ is a bijection,

(b) as a K-algebra, R is finitely generated.

For the moment we shall postpone the justification of the choice of language, and instead make an adjustment to the terminology. Suppose that we have the above situation. We may then use R to impose a certain structure on V. More precisely R may be employed to derive certain distinguished subsets of V, and in the case where we have a second affine set (W, S) to determine the significant mappings of V into W. In practice we shall modify the language and say that V itself is an affine set and we shall refer to R as its coordinate ring. The elements of V are usually referred to as points and from now on the coordinate ring of V will be denoted by $K[V]$. Note that, as x varies in V, M_x ranges over all the rational maximal ideals of $K[V]$ giving each one once and once only. The $K[V]$-topology on V will be called the affine topology.

Theorem 12. Let V be an affine set and let $x \in V$. Then $\{x\}$ is a closed subset of V. Also if $x, y \in V$, then $x \neq y$ if and only if there exists $f \in K[V]$ such that $f(x) = 0$, $f(y) \neq 0$.

Proof. By Lemma 1, $C_V(M_x) = \{x\}$. This establishes the first assertion and the second follows from the same lemma.

We add some further simple consequences of the definition of an affine set.

Lemma 4. Let V be an affine set. Then there exists a finite set $u_1, u_2, \ldots, u_n$ of elements of the coordinate ring $K[V]$ such that $K[V] = K[u_1, u_2, \ldots, u_n]$. Hence there is a surjective homomorphism

$$K[X_1, X_2, \ldots, X_n] \to K[V]$$

of K-algebras, where $X_1, X_2, \ldots, X_n$ are indeterminates that are mapped into $u_1, u_2, \ldots, u_n$ respectively.

Proof. By the definition of an affine set, $K[V]$ is a finitely generated K-algebra. The lemma follows by taking $u_1, u_2, \ldots, u_n$ to be any system of generators.

Theorem 13. Let V be an affine set and let $f_1, f_2, \ldots, f_r$ belong to $K[V]$. Then the following statements are equivalent:

(a) $f_1, f_2, \ldots, f_r$ are linearly independent over K;

(b) there exist $y_1, y_2, \ldots, y_r$ in V such that the determinant $|f_i(y_j)|$ is non-zero.

Proof. Assume (a). We can certainly find y_1 so that $f_1(y_1) \neq 0$. We can then find y_2 with the property that

$$\begin{vmatrix} f_1(y_1) & f_1(y_2) \\ f_2(y_1) & f_2(y_2) \end{vmatrix} \neq 0.$$

And so on. Thus (a) implies (b) and the converse is obvious.

Let V be an affine set with coordinate ring $K[V]$. If now $\mathfrak{A}$ is an ideal of $K[V]$, then $\mathrm{Rad}_V(\mathfrak{A})$, as defined in section (2.3), is the intersection of all the rational maximal ideals of $K[V]$ that contain $\mathfrak{A}$.

This is a convenient point at which to draw attention to what happens if K should be algebraically closed. Suppose that this is the case and let $X_1, X_2, \ldots, X_n$ be indeterminates. We have already observed that in the situation under consideration every maximal ideal of $K[X_1, X_2, \ldots, X_n]$ is K-rational. But, whatever the field K, the ring $K[X_1, X_2, \ldots, X_n]$ has the property that for any ideal $\mathfrak{A}$ the intersection of all the maximal ideals containing $\mathfrak{A}$ is $\mathrm{Rad}\,\mathfrak{A}$,† i.e. the set of elements for which some power belongs to the ideal. A ring with this property is sometimes called a Hilbert ring because of the intimate connection with Hilbert's Zeros Theorem. The same rings are also known as Jacobson rings because they are characterized by the fact that in every homomorphic image the nil-

† See, for example, [(9) Theorem 1, Cor. 1, p. 279].

radical is the same as the Jacobson radical. Because of the geometrical association we shall use the former name. In fact the general theory of these rings was discovered independently by O. Goldman (5) and W. Krull (7).

It is clear that any homomorphic image of a Hilbert ring is itself a Hilbert ring. Consequently, in view of Lemma 4, *if* V *is an affine set defined over an algebraically closed field* K, *then* $K[V]$ *is a Hilbert ring and all its maximal ideals are* K*-rational. Consequently for any ideal* $\mathfrak{A}$*, of* $K[V]$, $\mathrm{Rad}_V(\mathfrak{A}) = \mathrm{Rad}(\mathfrak{A})$. The next theorem also illustrates a special feature pertaining to the case where the ground field is algebraically closed.

Theorem 14. *Let* V *be an affine set defined over an algebraically closed field* K *and let* $f \in K[V]$. *Then* f *is a unit in the algebra* $K[V]$ *if and only if* f *does not vanish at any point of* V.

Proof. To say that f does not vanish at any point of V is equivalent to saying that f does not belong to any of the ideals M_x $(x \in V)$. But, because K is algebraically closed, this is equivalent to saying that f is not in any maximal ideal and so the theorem follows.

For the remainder of Chapter 2 we revert to our original assumptio[n], namely that K is an *arbitrary* field. In the next theorem S denotes a unitary, associative and commutative K-algebra.

Theorem 15. *Suppose that* S *is finitely generated as a* K*-algebra and let* $(\mathfrak{M}_K(S), \bar{S})$ *be its derived function algebra. Then* $\mathfrak{M}_K(S)$ *is an affine set (defined over* K*) having* $\bar{S}$ *as its coordinate ring.*

Proof. Since $\bar{S}$ is a homomorphic image of S it is finitely genera[-]ted as a K-algebra. Consequently we have only to establish that there is a bijection, of the right kind, between $\mathfrak{M}_K(S)$ and the set of K-rational maximal ideals of $\bar{S}$. However this has already been done in the discussi[on] which follows Lemma 2.

We shall now characterize (up to isomorphism) the K-algebras whic[h] are the coordinate rings of affine sets. To this end we make the

Definition. An 'affine K-algebra' is a K-algebra that is isomorphic to the coordinate ring of some affine set defined over K.

Theorem 16. Let S be a unitary, associative and commutative K-algebra. Then the following statements are equivalent:

(a) S is finitely generated (as a K-algebra) and is rationally reduced;

(b) S is an affine K-algebra.

Proof. Assume (a) and let us use the notation of Theorem 15. By Lemma 2, S and $\overline{S}$ are isomorphic K-algebras and, by Theorem 15, $\overline{S}$ is an affine algebra. This shows that (a) implies (b). The converse, of course, is trivial.

Corollary. The polynomial ring $K[X_1, X_2, \ldots, X_n]$ is an affine K-algebra if and only if K is an infinite field.

This is now immediate in view of Theorem 3.

Let V be an affine set and U an arbitrary subset of V. By restriction each f in K[V] induces a mapping $f^* : U \to K$, and the set of all these restrictions forms a K-algebra S say. In fact (U, S) is a function algebra over K, the mapping $K[V] \to S$ in which $f \mapsto f^*$ is a surjective homomorphism of K-algebras, and the kernel of the homomorphism is $I_V(U)$.

Let $y \in V$. By (2.3.10), $I_V(U) \subseteq M_y$ if and only if $y \in C_V(I_V(U))$; also, by Theorem 9, $C_V(I_V(U))$ is the closure, $\overline{U}$, of U in V. It follows that as y varies in $\overline{U}$, M_yS ranges over the K-rational maximal ideals of S giving rise to each one exactly once.

Since S is a homomorphic image of K[V], it is a finitely generated K-algebra. Let $x \in U$. Then

$$\{f^* \mid f \in K[V] \text{ and } f^*(x) = 0\}$$

is just M_xS. Thus different points of U give rise to different K-rational maximal ideals of S, but unless $\overline{U} = U$ not all the rational maximal ideals of S are obtained. This proves

Theorem 17. *Let* V *be an affine set and* U *a subset of* V. *Let* S *be the* K-*algebra consisting of the restrictions, to* U, *of the functions which make up* K[V]. *Then in order that* U *should be an affine set with* S *as its coordinate ring it is necessary and sufficient that* U *be a closed subset of* V.

From now on, whenever we have a closed subset of an affine set we shall regard the subset itself as being an affine set with the coordinate ring as described in Theorem 17.

Corollary. *Let* U *be a closed subset of the affine set* V. *Then* U *can be regarded as an affine set and as such it has an affine topology. Th coincides with the subspace topology.*

The assertion is clear because the principal loci on U are simply the contractions of the principal loci on V.

2.5 Examples of affine sets

In this section we assemble some examples of affine sets. They wi all be defined over the field K. We recall that if V is a set, then $\mathfrak{F}_K(V)$ denotes the K-algebra formed by all K-valued functions on V.

Example 1. Let V be an affine set with only a finite number of points. Let these be denoted by $x_1, x_2, \ldots, x_m$. Suppose $2 \le i \le m$. By Theorem 12, there exists $f_i \in K[V]$ such that $f_i(x_1) \neq 0$, $f_i(x_i) = 0$. Put $g_1 = cf_2f_3 \ldots f_m$, where $c \in K$. Then if c is suitably chosen we have $g_1(x_1) = 1$ and $g_1(x_i) = 0$ for $2 \le i \le m$. Of course $g_1 \in K[V]$. In fact we can find $g_1, g_2, \ldots, g_m$ in K[V] so that $g_i(x_j) = \delta_{ij}$. Finally let $k_1, k_2, \ldots, k_m$ belong to K and put $g = k_1g_1 + k_2g_2 + \ldots + k_mg_m$. Then $g(x_i) = k_i$ for $1 \le i \le m$ and we have proved

Theorem 18. *Let* V *be a finite affine set. Then* $K[V] = \mathfrak{F}_K(V)$.

Conversely if V consists of m distinct elements $x_1, x_2, \ldots, x_m$ then V is an affine set with $\mathfrak{F}_K(V)$ as its coordinate ring. For put

$$M_i = \{f \mid f \in \mathfrak{F}_K(V) \text{ and } f(x_i) = 0\}$$

Then $M_1, M_2, \ldots, M_m$ are distinct rational maximal ideals of $\mathfrak{F}_K(V)$. Further if M is a rational maximal ideal, then, because $M_1 M_2 \ldots M_m = 0$, we have $M = M_i$ for some i. Of course, $\mathfrak{F}_K(V)$ is a finitely generated K-algebra.

Example 2. Let V be an n-dimensional vector space over K and let $V^* = \mathrm{Hom}_K(V, K)$. Thus V^* is the K-space of <u>linear forms</u> on V or (as it is sometimes known) the <u>dual space.</u> Let $\xi \in V^*$, i.e. let ξ be a typical linear form. Since $\xi : V \to K$ it belongs to $\mathfrak{F}_K(V)$. Denote by $K[V]$ the subalgebra of the K-algebra $\mathfrak{F}_K(V)$ which the linear forms generate. We claim that V <u>is an affine set with coordinate ring</u> $K[V]$.

To see this let $e_1, e_2, \ldots, e_n$ be a base for the K-space V. Then V^* has a base $\xi_1, \xi_2, \ldots, \xi_n$, where $\xi_i(e_j) = \delta_{ij}$. This is the so-called <u>dual base.</u> Evidently $K[V] = K[\xi_1, \xi_2, \ldots, \xi_n]$.

Let $X_1, X_2, \ldots, X_n$ be indeterminates. We have a bijection between K^n and V in which $(\alpha_1, \alpha_2, \ldots, \alpha_n)$ in K^n is matched with $\alpha_1 e_1 + \alpha_2 e_2 + \ldots + \alpha_n e_n$. Let $\phi(X_1, X_2, \ldots, X_n)$ belong to $K[X_1, X_2, \ldots, X_n]$ and put $\xi = \phi(\xi_1, \xi_2, \ldots, \xi_n)$ so that ξ is a typical member of $K[V]$. If now $v = \alpha_1 e_1 + \alpha_2 e_2 + \ldots + \alpha_n e_n$ belongs to V and we identify V with K^n, then $\xi(v) = \phi(\alpha_1, \alpha_2, \ldots, \alpha_n)$. Thus $\xi : V \to K$ is just $\phi(X_1, X_2, \ldots, X_n)$ considered as a function from K^n to V. Accordingly $(V, K[V])$ has been identified with the derived function algebra of $K[X_1, X_2, \ldots, X_n]$. Consequently our claim follows from Theorem 15.

The example at the end of section (2.2) shows that the mapping

$$K[X_1, X_2, \ldots, X_n] \to K[V]$$

in which $\phi(X_1, \ldots, X_n) \mapsto \phi(\xi_1, \ldots, \xi_n)$ is an isomorphism of K-algebras if and only if K is infinite. We therefore obtain

Theorem 19. <u>Let</u> V <u>be an</u> n-<u>dimensional vector space over</u> K <u>and</u> $\xi_1, \xi_2, \ldots, \xi_n$ <u>any base of</u> $V^* = \mathrm{Hom}_K(V, K)$. <u>Then</u> V <u>is an affine set with</u> $K[\xi_1, \xi_2, \ldots, \xi_n]$ <u>as its coordinate ring. Moreover</u>

$\xi_1, \xi_2, \ldots, \xi_n$ _are algebraically independent over_ K _if and only if_ K _is infinite. If_ K _is finite, then_ V _is a finite set and_
$K[\xi_1, \xi_2, \ldots, \xi_n] = \mathfrak{F}_K(V)$.

The members of $K[V]$ are often referred to as _polynomial functions_ on V.

Example 3. This is essentially the same as Example 2 but the restatement involved helps to explain our terminology. K^n is an n-dimensional vector space over K and it has the natural base $e_1, e_2, \ldots, e_n$, where $e_i = (0, \ldots, 1, \ldots, 0)$ with the 1 occurring in the i-th position. Let $\overline{X}_i : K^n \to K$ be the i-th _coordinate function_ so that

$$\overline{X}_i(\alpha_1, \alpha_2, \ldots, \alpha_n) = \alpha_i .$$

Then $\overline{X}_1, \overline{X}_2, \ldots, \overline{X}_n$ is the base of $\mathrm{Hom}_K(K^n, K)$ that is dual to $e_1, e_2, \ldots, e_n$. Accordingly K^n is an affine set with $K[\overline{X}_1, \overline{X}_2, \ldots, \overline{X}_n]$ as its coordinate ring. We refer to this affine set as _affine_ n-_space._ Of course

$$K[\overline{X}_1, \overline{X}_2, \ldots, \overline{X}_n] = \overline{K[X_1, X_2, \ldots, X_n]},$$

where the right hand side consists of the members of $K[X_1, X_2, \ldots, X_n]$ regarded as functions from K^n to K, and the natural homomorphism

$$K[X_1, X_2, \ldots, X_n] \to \overline{K[X_1, X_2, \ldots, X_n]}$$

maps X_i into $\overline{X}_i$.

Let $\phi(X_1, X_2, \ldots, X_n) \in K[X_1, X_2, \ldots, X_n]$ and let $\overline{\phi}$ be the corresponding member of the coordinate ring of K^n. The principal locus determined by $\overline{\phi}$ is composed of the points $(\alpha_1, \alpha_2, \ldots, \alpha_n)$ of K^n that are zeros of $\phi(X_1, X_2, \ldots, X_n)$. It follows that the closed subsets of affine n-space are just those loci that are defined by algebraic equations in the manner that is familiar from elementary coordinate geometry. Of course when K is infinite we can regard $K[X_1, X_2, \ldots, X_n]$ itself as the coordinate ring of K^n.

Example 4. This is an important example of a way in which affine sets often arise. Let V be a given affine set and let $f \in K[V]$. Put

$$U = \{x \mid x \in V \text{ and } f(x) \neq 0\},$$

and observe that, because U is the complement of a principal locus, it is an <u>open</u> subset of V.

Each $g \in K[V]$ determines by restriction a mapping $g^* : U \to K$. If we denote the set of these restrictions by S, then S is a subalgebra of the K-algebra $\mathfrak{F}_K(U)$, and the mapping $g \mapsto g^*$ is a surjective homomorphism $K[V] \to S$ of K-algebras. Evidently (U, S) is a function algebra over K but, in view of Theorem 17, it need not be an affine set even though, as is clear, S is a finitely generated K-algebra.

The restriction f^*, of f, vanishes nowhere on U and so we have a mapping $U \to K$ in which $x \mapsto 1/f(x)$. This mapping will be denoted by $1/f^*$. The smallest subalgebra of $\mathfrak{F}_K(U)$ containing S and this new function is $S[1/f^*]$. This too is finitely generated as a K-algebra. Also $(U, S[1/f^*])$ is a function algebra over K.

Theorem 20. <u>Let the notation be as above. Then</u> U <u>is an affine set with</u> $S[1/f^*]$ <u>as its coordinate ring.</u>

Proof. Let $x, y \in U$ with $x \neq y$. By Theorem 12, there exists $g \in K[V]$ such that $g(x) = 0$, $g(y) \neq 0$. The restriction g^*, of g to U, belongs to $S[1/f^*]$ and, of course, $g^*(x) = 0$, $g^*(y) \neq 0$. This shows that different elements of U give rise to different maximal ideals of $S[1/f^*]$.

Let N be a rational maximal ideal of $S[1/f^*]$. It follows (Theorem 1) that $N \cap S$ is a rational maximal ideal of S and therefore there exists $z \in V$ such that $M_z S = N \cap S$. Since f^* is a unit in $S[1/f^*]$, we see that $f^* \notin N$ and therefore $f \notin M_z$. This shows that $z \in U$.

Now assume that $\psi \in N$. For some $\nu > 0$ we have $(f^*)^\nu \psi \in M_z S$. Consequently $\psi(z) = 0$ and therefore

$$N \subseteq \{\omega \mid \omega \in S[1/f^*] \text{ and } \omega(z) = 0\}.$$

However there must be equality because N is a maximal ideal. Thus we have the desired bijection between U and the set of rational maximal ideals of $S[1/f^*]$. The theorem is therefore proved.

An important simplification takes place when f is a _non-zero-divisor_ in $K[V]$. Assume that

$$g \in \mathrm{Ker}(K[V] \to S).$$

Then fg vanishes everywhere on V and hence $g = 0$. Thus $K[V] \to S$ is an isomorphism of K-algebras and it leads to an isomorphism

$$K[V]\left[\frac{1}{f}\right] \approx S\left[\frac{1}{f^*}\right]$$

on the basis of which we can make the identification

$$K[U] = K[V]\left[\frac{1}{f}\right]. \tag{2.5.1}$$

It is clear how the members of the right hand side are to be regarded as K-valued functions defined on U. We also see that _if_ $K[V]$ _is an integral domain and_ $f \neq 0$, _then_ $K[U]$ _is also an integral domain._

Example 5. Let V be an n-dimensional vector space over K and put $E = \mathrm{End}_K(V)$ so that E consists of all the K-linear mappings of V into itself.† E is a unitary, associative K algebra. In particular it is a vector space over K and as such its dimension is n^2. Accordingly, by Example 2, there is a natural way in which E can be regarded as an affine set.

We now define a mapping

$$D : E \to K \tag{2.5.2}$$

by means of

$$D(g) = \text{the determinant of } g \tag{2.5.3}$$

for each $g \in E$. Clearly D belongs to the coordinate ring, $K[E]$, of E. We shall apply the construction described in Example 4 to this situation.

To this end we put

† See section (1.3) Example 3.

$$GL(V) = \{g \mid g \in E \text{ and } D(g) \neq 0\}. \tag{2.5.4}$$

Then $GL(V)$ consists of the units of the algebra $\mathrm{End}_K(V)$ and therefore it has a group structure. It is usually referred to as the general linear group of V.

For the moment our main concern is to note that the considerations put forward under Example 4 enable us to regard $GL(V)$ as an affine set. In fact if S is the K-algebra obtained by restricting the members of $K[E]$ to $GL(V)$, then the coordinate ring of $GL(V)$ is given by

$$K[GL(V)] = S\left[\frac{1}{D^*}\right] , \tag{2.5.5}$$

where D^* denotes the restriction of D.

For the remainder of section (2.5) we shall assume that K is an infinite field. Theorem 19 then shows that $K[E]$ is an integral domain and therefore, by (2.5.1), we may write

$$K[GL(V)] = K[E]\left[\frac{1}{D}\right] . \tag{2.5.6}$$

Indeed we can find a base f_{ij} $(1 \leq i \leq n,\ 1 \leq j \leq n)$ for E with the following property: if $f = \Sigma\Sigma a_{ij} f_{ij}$, where $a_{ij} \in K$, then $D(f)$ is the determinant of the matrix $\|a_{ij}\|$. This enables us to identify $K[E]$ with $K[X_{11}, X_{12}, \ldots, X_{nn}]$, where the X_{ij} are indeterminates, and then D in $K[E]$ becomes identified with the polynomial $\mathrm{Det}\|X_{ij}\|$. Thus to sum up: when K is an infinite field

$$K[GL(V)] = K[X_{11}, X_{12}, \ldots, X_{nn}]\left[\frac{1}{\mathrm{Det}\|X_{ij}\|}\right] . \tag{2.5.7}$$

In this context if $f = \Sigma\Sigma a_{ij} f_{ij}$ belongs to $GL(V)$ and we wish to evaluate a member of the right hand side of (2.5.7) at f, then we do this by assigning to X_{ij} the value a_{ij}.

2.6 Morphisms of affine sets

Throughout section (2.6) we shall use V and W to denote given affine sets that are defined over K. Let $x \in V$ and $y \in W$. It will be convenient to use M_x respectively N_y to denote the corresponding rational maximal ideal of K[V] respectively K[W].

Suppose that we have a mapping $\phi : V \to W$. For $g \in K[W]$ put

$$\phi^*(g) = g \circ \phi. \tag{2.6.1}$$

Then $\phi^*(g) : V \to K$ and, with a self-explanatory notation,

$$\phi^*(g_1 + g_2) = \phi^*(g_1) + \phi^*(g_2),$$
$$\phi^*(g_1 g_2) = \phi^*(g_1)\phi^*(g_2),$$
$$\phi^*(kg) = k\phi^*(g).$$

Of course ϕ^* preserves constant mappings.

Definition. The mapping $\phi : V \to W$ is called a 'K-morphism' of affine sets if $\phi^*(g) \in K[V]$ for all $g \in K[W]$.

Suppose that $\phi : V \to W$ is a K-morphism. This induces a homomorphism

$$\phi^* : K[W] \to K[V] \tag{2.6.2}$$

of K-algebras in which $g \mapsto g \circ \phi$. It follows that if $x \in V$, then

$$\phi^{*-1}(M_x) = N_{\phi(x)} \tag{2.6.3}$$

Consequently _if_ $\phi_1, \phi_2 : V \to W$ _are_ K-_morphisms and_ $\phi_1^* = \phi_2^*$, _then_ $\phi_1 = \phi_2$.

On the other hand suppose that we start with a homomorphism $\omega : K[W] \to K[V]$ of K-algebras. By Theorem 1 and the definition of an affine set, there is a uniquely determined mapping $\psi : V \to W$ such that

$$\omega^{-1}(M_x) = N_{\psi(x)}$$

for all $x \in V$. If now $g \in K[W]$, then $\omega(g - (\omega(g))(x))$ belongs to M_x and therefore $g - (\omega(g))(x)$ is in $N_{\psi(x)}$. Thus

$$(\psi^*(g))(x) = g(\psi(x)) = (\omega(g))(x)$$

from which it follows that $\psi^*(g) = \omega(g) \in K[V]$. We now see that $\psi : V \to W$

is a K-morphism and $\psi^* = \omega$. Moreover by combining our observations we obtain the important

Theorem 21. There is a natural bijection between the K-morphisms $V \to W$ and the homomorphisms $K[W] \to K[V]$ of K-algebras. If $\phi : V \to W$ is a K-morphism of affine sets, then the algebra homomorphism associated with it is $\phi^* : K[W] \to K[V]$, where this is defined in the manner explained above.

We shall make a fairly deep study of K-morphisms at a later stage. For the moment we shall content ourselves with some simple observations. First of all the identity mapping of V is a K-morphism and it is associated with the identity homomorphism of $K[V]$. Next suppose that U is a third affine set and that $\phi : V \to W$ and $\chi : W \to U$ are K-morphisms. We observe, for future reference, that $\chi \circ \phi : V \to U$ is a K-morphism and that

$$(\chi \circ \phi)^* = \phi^* \circ \chi^*. \tag{2.6.4}$$

Our aim is to study affine sets and their morphisms. It will now be clear that this is virtually equivalent to studying affine algebras and their homomorphisms. Thus we could, if we wished, restrict our attention to the study of algebras. This indeed would have substantial advantages including the possibility of achieving great generality. However we shall not do this because it would sever the connections of the subject with geometry and, in a certain sense, it would stand the subject on its head. During the twentieth century Geometry has been the victim of a takeover bid by Algebra and a very useful way of looking at certain questions has been lost to some mathematicians. For once generality will be asked to give way to historical considerations.

Returning to the main discussion, let $\phi : V \to W$ be a K-morphism of affine sets.

Definition. We say that ϕ is a 'K-isomorphism' of V on to W if it is a bijection and has the property that $\phi^{-1} : W \to V$ is also a K-morphism.

Evidently $\phi : V \to W$ is a K-isomorphism if and only if $\phi^* : K[W] \to K[V]$ is an isomorphism of K-algebras. Furthermore when this is the case we have

$$(\phi^{-1})^* = (\phi^*)^{-1} .$$

Example. It is important to note that a K-morphism can be a bijection without being a K-isomorphism. The following example illustrates this point.

Let K be an algebraically closed field of characteristic p, where p is a prime. Since K is infinite, if we regard K itself as affine 1-space, then its coordinate ring may be taken to be K[X], where X is an indeterminate. Define $\phi : K \to K$ by $\phi(\alpha) = \alpha^p$. This is a K-morphism and $\phi^* : K[X] \to K[X]$ maps $g(X) \in K[X]$ into $g(X^p)$. Consequently ϕ is not a K-isomorphism. However it certainly is bijective.

The next theorem provides the justification for the name _affine set._ Before proceeding to prove it we observe that _any_ bijection between two _finite_ affine sets is a K-isomorphism. This follows from Theorem 18.

Theorem 22. _Let V be an affine set defined over K. If n is a sufficiently large positive integer, then V is K-isomorphic to a closed subset of_ K^n.

Proof. By Lemma 4, we can construct a surjective homomorphism

$$K[X_1, X_2, \ldots, X_n] \to K[V] \tag{2.6.5}$$

of K-algebras, where $X_1, X_2, \ldots, X_n$ are indeterminates. If K is a finite field, then $K[X_1, X_2, \ldots, X_n]$, and hence also K[V], has only a finite number of rational maximal ideals, and therefore V has only a finite number of points. But in this case what we have to prove is trivial.

From here on we assume that K is _infinite._ Let $\mathfrak{A}$ be the kernel of (2.6.5) and U the locus† of $\mathfrak{A}$. Then U is a closed subset of K^n. Because K[V] is rationally reduced, $\mathfrak{A}$ is the intersection of all the

† We regard $K[X_1, X_2, \ldots, X_n]$ as the coordinate ring of K^n.

rational maximal ideals that contain it. Accordingly, by (2. 3. 15), $I_{K^n}(U) = \mathfrak{A}$ and therefore

$$K[U] \approx K[X_1, X_2, \ldots, X_n]/\mathfrak{A} \approx K[V]$$

these being isomorphisms of K-algebras. It follows U and V are K-isomorphic.

Corollary.[‡] <u>If K[V] can be generated, as a K-algebra, by $n \geq 1$ elements, then n is a large enough integer for the purposes of the theorem.</u>

The next result shows that K-morphisms are <u>continuous.</u>

Theorem 23. <u>Let $\phi : V \to W$ be a K-morphism of affine sets and let each of V and W be endowed with its affine topology. Then ϕ is continuous.</u>

Proof. Let $g \in K[W]$ and put

$$C = \{y \mid y \in W \text{ and } g(y) = 0\}.$$

Since every closed subset of W is an intersection of principal loci, it will suffice to show that $\phi^{-1}(C)$ is a principal locus in V. But this is clear because, if $f = g \circ \phi$, then $f \in K[V]$ and

$$\phi^{-1}(C) = \{x \mid x \in V \text{ and } f(x) = 0\}.$$

2. 7 Products of function algebras

Throughout section (2. 7) we shall use (V, R) and (W, S) to denote function algebras over K. Let $f \in R$, $g \in S$ and define

$$f \vee g : V \times W \to K \tag{2. 7. 1}$$

by

$$(f \vee g)(x, y) = f(x)g(y). \tag{2. 7. 2}$$

---

‡ This holds for arbitrary K.

The functions $f \vee g$ will generate a subalgebra of the K-algebra $\mathfrak{F}_K(V \times W)$. This subalgebra will be denoted by $R \times_K S$. Accordingly $(V \times W, R \times_K S)$ is a new function algebra over K. It will be called the product of (V, R) and (W, S). Note that a typical element of $R \times_K S$ can be written in the form

$$f_1 \vee g_1 + f_2 \vee g_2 + \ldots + f_m \vee g_m,$$

where $f_i \in R$ and $g_i \in S$. Note also that

$$(f_1 \vee g_1)(f_2 \vee g_2) = f_1 f_2 \vee g_1 g_2. \tag{2.7.3}$$

Again it is easy to check that there is a K-linear mapping

$$R \otimes_K S \to R \times_K S$$

in which $f \otimes g \mapsto f \vee g$, and then a further simple verification shows this to be a surjective homomorphism of K-algebras. However much more is true as is shown by

Theorem 24. The K-linear mapping $R \otimes_K S \to R \times_K S$, in which $f \otimes g$ is mapped into $f \vee g$, is an isomorphism of K-algebras.

Proof. Let $f_1 \otimes g_1 + f_2 \otimes g_2 + \ldots + f_m \otimes g_m$ belong to the kern of the mapping. We have to show that this is zero. Without loss of generality we may assume that $g_1, g_2, \ldots, g_m$ are linearly independen over K.

Let $x \in V$ and put $g = f_1(x)g_1 + f_2(x)g_2 + \ldots + f_m(x)g_m$. Then g vanishes identically on W and therefore

$$f_1(x)g_1 + f_2(x)g_2 + \ldots + f_m(x)g_m = 0.$$

It follows that $f_i(x) = 0$ for $i = 1, 2, \ldots, m$. This shows that $f_1, f_2, \ldots, f_m$ are all zero and the theorem is established.

When it is convenient we shall use Theorem 24 to identify $R \otimes_K S$ with $R \times_K S$. Let $\mathfrak{A}$ be an ideal of R and $\mathfrak{B}$ an ideal of S. Then†

† See (1.5.4).

$[\mathfrak{A}, \mathfrak{B}]$ is an ideal of $R \otimes_K S = R \times_K S$. In fact a typical element of $[\mathfrak{A}, \mathfrak{B}]$ can be written in the form

$$f_1 \vee g_1 + f_2 \vee g_2 + \ldots + f_m \vee g_m,$$

where for each i $(1 \leq i \leq m)$ either $f_i \in \mathfrak{A}$ and $g_i \in S$ or $f_i \in R$ and $g_i \in \mathfrak{B}$.

Let $x \in V$ and $y \in W$. The former of these determines a rational maximal ideal M_x of R and the latter a rational maximal ideal N_y of S. By Theorem 5, $[M_x, N_y]$ is a rational maximal ideal of $R \otimes_K S = R \times_K S$. Suppose that $\phi \in [M_x, N_y] \subseteq R \otimes_K S$. The remarks of the last paragraph show that $\phi(x, y) = 0$. Accordingly we obtain

Lemma 5. Let the notation be as above. Then

$$[M_x, N_y] = \{\phi \mid \phi \in R \times_K S \text{ and } \phi(x, y) = 0\}$$

for all $x \in V$ and $y \in W$.

We conclude this section by applying our observations to prove

Theorem 25. Let R' and S' be rationally reduced K-algebras.† Then $R' \otimes_K S'$ is also a rationally reduced K-algebra.

Proof. Let (V, R) respectively (W, S) be the derived function algebra of R' respectively S'. By Lemma 2, R' is isomorphic to R and S' is isomorphic to S. Consequently, by Theorem 24, we have isomorphisms

$$R' \otimes_K S' \approx R \otimes_K S \approx R \times_K S$$

of K-algebras, and $R \times_K S$ is rationally reduced because $(V \times W, R \times_K S)$ is a function algebra (see Theorem 6).

† It is assumed that R' and S' are unitary, associative, and commutative.

2.8 Products of affine sets

Throughout section (2.8) the letters V and W denote given affine sets defined over K. Also if $x \in V$ and $y \in W$, then M_x and N_y denote the corresponding rational maximal ideals of $K[V]$ and $K[W]$ respectively.

By Theorem 24, there is an isomorphism

$$K[V] \otimes_K K[W] \approx K[V] \times_K K[W] \qquad (2.8.1)$$

of K-algebras in which $f \otimes g$ is matched with $f \vee g$. Let us use (2.8.1) to identify the two algebras. Then, by Lemma 5,

$$[M_x, N_y] = \{\phi \mid \phi \in K[V] \times_K K[W] \text{ and } \phi(x, y) = 0\}. \qquad (2.8.2)$$

Also, by section (1.5), $K[V] \times_K K[W]$ is a finitely generated K-algebra. The next result is now an immediate consequence of Theorem 5 and the fact that V and W are affine sets.

Theorem 26. _Let_ V _and_ W _be affine sets defined over_ K. _Then the cartesian product_ $V \times W$ _has a natural structure as an affine set with_ $K[V \times W] = K[V] \times_K K[W]$.

Since $V \times W$ is an affine set it has an affine topology. It is important to note that this is not the product of the affine topology of V with that of W. This is illustrated by the following example.

Let K be an _infinite_ field and X, Y indeterminates. Take $V = K$, $W = K$ and identify their coordinate rings with $K[X]$ and $K[Y]$. The coordinate ring of $K \times K$ is then $K[X] \otimes K[Y] = K[X, Y]$ which shows that the product of K with itself is the same as affine 2-space. Put

$$\Delta = \{(x, y) \mid (x, y) \in K \times K \text{ and } x = y\}.$$

Then Δ is a closed subset of $K \times K$ when $K \times K$ has its affine topology.

Assume that O_1 and O_2 are non-empty open subsets of K. Then, as we saw in section (2.3), their complements are finite. Consequently there exists $\alpha \in K$ such that $(\alpha, \alpha) \in O_1 \times O_2$. Accordingly when $K \times K$ is endowed with the product topology every non-empty open set meets Δ.

Hence the product topology on $K \times K$ is different from the affine topology.

We continue now with the general discussion. Let A be a closed subset of V and B a closed subset of W. By identifying $K[V \times W]$ with $K[V] \otimes K[W]$ we can regard $[I_V(A), I_W(B)]$ as an ideal of the former. The locus of this ideal is $A \times B$ and therefore $A \times B$ _is a closed subset of_ $V \times W$. Note that if P and Q are open subsets of V and W respectively, then $P \times W$ and $V \times Q$ are open subsets of $V \times W$ because they are the complements of $(V \backslash P) \times W$ and $V \times (W \backslash Q)$ respectively. Accordingly $(P \times W) \cap (V \times Q)$, that is to say $P \times Q$, is an open subset of $V \times W$.

Since $A \times B$ is a closed subset of $V \times W$ we have two ways of regarding $A \times B$ as an affine set. First of all we may regard A and B as affine sets and then $A \times B$ is an affine set in its own right. Secondly $A \times B$ is a closed subset of $V \times W$ and as such inherits an affine structure. However, as the next theorem shows, there is no danger of ambiguity.

Theorem 27. _Let_ A _be a closed subset of the affine set_ V _and_ B _a closed subset of the affine set_ W. _Then the two ways of regarding_ $A \times B$ _as an affine set (see above) yield the same coordinate ring on_ $A \times B$.

Proof. Let $f_1, f_2, \ldots, f_s$ belong to $K[V]$ and $g_1, g_2, \ldots, g_s$ to $K[W]$. Denote by f_i^* respectively g_i^* the restriction of f_i respectively g_i to A respectively B. The theorem now follows from the observation that

$$f_1^* \vee g_1^* + f_2^* \vee g_2^* + \ldots + f_s^* \vee g_s^*$$

is the restriction of $f_1 \vee g_1 + f_2 \vee g_2 + \ldots + f_s \vee g_s$ to $A \times B$.

So far we have only considered products of _two_ affine sets but the concept extends easily to arbitrary finite products. Since no essentially new ideas are involved, we shall deal with the extension rather briefly.

Let $V_1, V_2, \ldots, V_s$ be affine sets defined over K and let $f_i \in K[V_i]$ for $i = 1, 2, \ldots, s$. From these we construct a function

$$V_1 \times V_2 \times \ldots \times V_s \to K$$

which is denoted by $f_1 \vee f_2 \vee \ldots \vee f_s$ and which is such that

$$(x_1, x_2, \ldots, x_s) \mapsto f_1(x_1) f_2(x_2) \ldots f_s(x_s).$$

These functions generate a subalgebra of the K-algebra $\mathfrak{F}_K(V_1 \times V_2 \times \ldots \times V_s)$. Denote the subalgebra by

$$K[V_1] \times_K K[V_2] \times_K \ldots \times_K K[V_s].$$

An easy argument using induction now shows that $V_1 \times V_2 \times \ldots \times V_s$ may be regarded as an affine set whose coordinate ring is given by

$$K[V_1 \times V_2 \times \ldots \times V_s] = K[V_1] \times_K K[V_2] \times_K \ldots \times_K K[V_s].$$

It should be noted that, when convenient,

$$V_1 \times V_2 \times \ldots \times V_s \text{ and } (V_1 \times V_2 \times \ldots \times V_m) \times (V_{m+1} \times V_{m+2} \times \ldots \times V_s)$$

may be identified in the obvious way. Also if $\{i_1, i_2, \ldots, i_s\}$ is a permutation of $\{1, 2, \ldots, s\}$, then the canonical bijection

$$V_1 \times V_2 \times \ldots \times V_s \overset{\sim}{\to} V_{i_1} \times V_{i_2} \times \ldots \times V_{i_s}$$

is a K-isomorphism.

2.9 Some standard morphisms

Certain operations with affine sets lead to mappings and it is important to know when these are morphisms. In this section we shall make a useful collection of results concerning such situations. To avoid repetition we record at the outset that <u>all affine sets considered in section (2.9) are to be understood as being defined over</u> K. It will be remembered from section (2.6) that an identity mapping of an affine set is a K-isomorphism and that the result of applying two K-morphisms in succession is another K-morphism.

Lemma 6. *Let V and W be affine sets and suppose that V is finite. Then every mapping $\phi : V \to W$ is a K-morphism.*

This follows from Theorem 18.

Lemma 7. *Let A be a closed subset of the affine set V and let $j : A \to V$ be the inclusion mapping. Then j is a K-morphism and the associated homomorphism $j^* : K[V] \to K[A]$ of K-algebras is surjective.*

This is clear because we obtain $K[A]$ by taking the functions which make up $K[V]$ and restricting their common domain to A.

Lemma 8. *Let $\phi : V \to W$ be a K-morphism of affine sets and let B be a closed subset of W. Suppose that $\phi(V) \subseteq B$. Then the induced mapping $V \to B$ is also a K-morphism of affine sets.*

For if $f \in K[W]$ restricts to f^* on B, then the mapping $x \mapsto f^*(\phi(x))$ of V into K is just $f \circ \phi$ and this belongs to $K[V]$.

Next suppose that $\phi : V \to V'$ and $\psi : W \to W'$ are mappings and define

$$\phi \times \psi : V \times W \to V' \times W' \tag{2.9.1}$$

by

$$(\phi \times \psi)(x, y) = (\phi(x), \psi(y)). \tag{2.9.2}$$

Lemma 9. *Let $\phi : V \to V'$ and $\psi : W \to W'$ be K-morphisms of affine sets. Then*

$$\phi \times \psi : V \times W \to V' \times W'$$

is a K-morphism as well.

This is clear because if $f' \in K[V']$ and $g' \in K[W']$, then

$$(f' \vee g') \circ (\phi \times \psi) = (f' \circ \phi) \vee (g' \circ \psi).$$

Lemma 10. *Let V and W be affine sets. Then the projections*

$$\pi_1 : V \times W \to V \text{ and } \pi_2 : V \times W \to W$$

are K-morphisms of affine sets.

Proof. We need only consider π_1. Let $f \in K[V]$. Then

$$f \circ \pi_1 = f \vee 1 \in K[V \times W]$$

and therefore π_1 is a K-morphism.

For Lemma 11 it will be assumed that V, W, W' are affine sets and that $\phi : V \to W$, $\phi' : V \to W'$ are mappings. Define $\psi : V \to W \times W'$ by $\psi(x) = (\phi(x), \phi'(x))$.

Lemma 11. With the above notation $\psi : V \to W \times W'$ is a K-morphism if and only if ϕ and ϕ' are K-morphisms.

Proof. Suppose that ϕ and ϕ' are K-morphisms and assume that $g \in K[W]$, $g' \in K[W']$. The fact that ψ is a K-morphism follows from the relation

$$(g \vee g') \circ \psi = (g \circ \phi)(g' \circ \phi').$$

On the other hand if ψ is a K-morphism, then, by combining ψ with the projections $V \times W \to V$ and $V \times W \to W$ and using Lemma 10, we see that ϕ and ϕ' are K-morphisms.

Next suppose that $x \in V$, $y \in W$ and define

$$\sigma_y : V \to V \times W \tag{2.9.3}$$
$$\tau_x : W \to V \times W \tag{2.9.4}$$

by $\sigma_y(x) = (x, y)$ and $\tau_x(y) = (x, y)$.

Lemma 12. Let V and W be affine sets and suppose that $x \in V$, $y \in W$. Then, with the above notation, σ_y and τ_x are K-morphisms.

Proof. Let $\omega \in K[V \times W]$ say

$$\omega = f_1 \vee g_1 + f_2 \vee g_2 + \ldots + f_s \vee g_s,$$

where $f_i \in K[V]$ and $g_i \in K[W]$. Then

$$\omega \circ \tau_x = f_1(x)g_1 + f_2(x)g_2 + \ldots + f_s(x)g_s \in K[W]$$

which shows that τ_x is a K-morphism. The proof that σ_y is a K-morphism is similar.

Let x belong to the affine set V and y to the affine set W. Then there is a bijection $V \approx V \times \{y\}$ in which ξ in V is matched with (ξ, y) in $V \times \{y\}$ and a bijection $W \approx \{x\} \times W$ which arises similarly. By Theorem 12, $\{x\}$ is a closed subset of V and $\{y\}$ a closed subset of W. Moreover $V \times \{y\}$ and $\{x\} \times W$ are affine sets.[†]

Theorem 28. *Let* V *and* W *be affine sets and suppose that* $x \in V$, $y \in W$. *Then the natural bijections* $V \approx V \times \{y\}$ *and* $W \approx \{x\} \times W$ *are K-isomorphisms of affine sets.*

Proof. By Lemma 12, $\sigma_y : V \to V \times W$ is a K-morphism and, of course, $V \times \{y\}$ is a closed subset of $V \times W$. Lemma 8 therefore shows that the bijection $V \to V \times \{y\}$ is a K-morphism. However the inverse mapping $V \times \{y\} \to V$ is a K-morphism because it is a projection. Accordingly $V \to V \times \{y\}$ is a K-isomorphism. The other assertion is proved similarly.

Corollary. *Let* V, W *be affine sets and let* A *respectively* B *be a subset of* V *respectively* W. *If now* $\bar{A}, \bar{B}$ *are the closures of* A, B *in* V, W *respectively, then the closure,* $\overline{A \times B}$ *say, of* $A \times B$ *in* $V \times W$ *is* $\bar{A} \times \bar{B}$.

Proof. Let $b \in B$. By Theorem 28, $\bar{A} \times \{b\}$ is the closure of $A \times \{b\}$ in $V \times \{b\}$. Since $(V \times \{b\}) \cap \overline{A \times B}$ contains $A \times \{b\}$, we see that $\bar{A} \times \{b\} \subseteq \overline{A \times B}$ and therefore $\bar{A} \times B \subseteq \overline{A \times B}$. Accordingly $\bar{A} \times B = A \times B$ and likewise $A \times \bar{B} = A \times B$. In the last equation replace A by $\bar{A}$. This yields

$$\bar{A} \times \bar{B} = \overline{\bar{A} \times \bar{B}} = \overline{\bar{A} \times B} = \overline{A \times B}$$

---

† In this connection the reader should recall Theorem 27.

as required.

We mention, in passing, a K-morphism which expresses the fact that an affine set V is, to a certain extent, <u>separated.</u>

Define $\delta : V \to V \times V$ by $\delta(x) = (x, x)$. This is a K-morphism by Lemma 11. The image of δ is a subset, Δ say, of $V \times V$ called its <u>diagonal.</u> By Theorem 12, this subset consists of the common zeros of the functions $f \vee 1 - 1 \vee f$ obtained by varying f in $K[V]$. Thus Δ is a <u>closed</u> subset of $V \times V$. Also δ induces a K-isomorphism, of V on to Δ, whose inverse is obtained by combining the inclusion mapping $\Delta \to V \times V$ with a projection.

The next result completes our collection of useful morphisms. It will be recalled that, in section (2.5), we showed that if V is a vector space over K of finite dimension, then V has a natural structure as an affine set.

Lemma 13. <u>Let V and V' be finite-dimensional vector spaces over K and let $\phi : V \to V'$ be a K-linear mapping. If now V and V' are regarded as affine sets, then ϕ is also a K-morphism of such sets.</u>

This is clear in view of the way in which polynomial functions, on V and V', are defined.

2.10 Enlargement of the ground field

In this section we shall examine the effect on an affine set of enlarging the ground field. Accordingly, throughout section (2.10), V and W will denote affine sets defined over K, and L will denote an extension field of K.

Suppose that $f \in K[V]$. Then $f : V \to K$ and therefore f determines a mapping of V into L. (This mapping will be denoted by the same letter.) Thus $K[V] \subseteq \mathfrak{F}_L(V)$. Denote by S the subalgebra of the L-algebra $\mathfrak{F}_L(V)$ that $K[V]$ generates. Then (V, S) is a function algebra over L. We must now identify S.

Let $\{\lambda_i\}_{i \in I}$ be a base for L over K and suppose that $\omega \in S$. Then ω can be represented in the form

$$\omega = \sum_{i \in I} \lambda_i f_i, \tag{2.10.1}$$

where $f_i \in K[V]$ and only finitely many of the f_i are non-zero. Let $x \in V$. Then $\omega(x) = 0$ if and only if $f_i(x) = 0$ for all i, and therefore $\omega = 0$ if and only if every f_i is zero. This shows that the representation of a member of S in the form (2.10.1) is unique and now it follows that

$$S = K[V]^L, \tag{2.10.2}$$

where the notation is the same as that employed in section (2.6). The results of that section show that S is a finitely generated L-algebra, whereas Theorem 4 Cor. 1 enables us to conclude that it is rationally reduced. Accordingly $S = K[V]^L$ is an affine L-algebra. Note that _if_ $f_1, f_2, \ldots, f_m$ _belong to_ $K[V]$ _and are linearly independent over_ K, _then they are linearly independent over_ L _when considered as elements of_ S.

Let $\omega \in S$ and let ω be represented as in (2.10.1). Further let $x \in V$ and denote by M_x the corresponding maximal ideal of $K[V]$. Then $\omega(x) = 0$ if and only if $f_i \in M_x$ for $i \in I$, that is if and only if $\omega \in M_x S$. Thus

$$M_x S = \{\omega \mid \omega \in S \text{ and } \omega(x) = 0\} \tag{2.10.3}$$

and

$$\omega(x) = M_x S\text{-residue of } \omega. \tag{2.10.4}$$

Since $M_x S \cap K[V] = M_x$, we conclude from (2.10.3) that different points of V give rise to different L-rational maximal ideals of S. Of course, usually not every L-rational maximal ideal of S will arise from a point of V. To these observations we add the obvious one that _if_ $\omega(x) = 0$ _for all_ $x \in V$, _then_ $\omega = 0$.

We now enlarge V to a new set $V^{(L)}$ which is put into a one-one correspondence with $\mathfrak{M}_L(S)$. At the same time we construct the bijection $V^{(L)} \to \mathfrak{M}_L(S)$ in such a way that x, in V, is matched with $M_x S$. Finally for $\omega \in S$ and $z \in V^{(L)}$ we define $\omega(z)$ to be the residue of ω with respect to the member of $\mathfrak{M}_L(S)$ that corresponds to z. The outcome is that $V^{(L)}$ becomes an affine set, defined over L, whose coordinate ring may be identified with S. Thus

$$L[V^{(L)}] \approx S = K[V]^L. \tag{2.10.5}$$

Of course $V^{(L)}$ is (essentially) uniquely determined by V and L. Also we obtain the isomorphism $L[V^{(L)}] \approx S$ of (2.10.5) by taking the L-valued functions that make up $L[V^{(L)}]$ and in each case restricting the domain to V.

Definition. $V^{(L)}$ is called the 'affine set obtained from V by enlarging the ground field from K to L'.

Note that if $f \in K[V] \subseteq K[V]^L$, then there is a unique member of $L[V^{(L)}]$ which becomes f when its domain is restricted to V. We call this the natural prolongation of f to $V^{(L)}$. Frequently the same letter is used for a member of $K[V]$ and its prolongation. Each member of $L[V^{(L)}]$ is a linear combination (with coefficients in L) of such prolongations.

Theorem 29. *Let the notation be as above. Then the affine topology on V is the same as the topology induced on it by the affine topology of $V^{(L)}$. Moreover the closure of V, in $V^{(L)}$, is $V^{(L)}$ itself.*

Proof. Let $\omega \in L[V^{(L)}]$ and write ω in the form

$$\omega = \lambda_1 f_1 + \lambda_2 f_2 + \ldots + \lambda_m f_m,$$

where $\lambda_1, \lambda_2, \ldots, \lambda_m$ belong to L and are linearly independent over K, and each f_i is (the prolongation of) a member of $K[V]$. Then the principal locus

$$\{z \mid z \in V^{(L)} \text{ and } \omega(z) = 0\}$$

intersects V in the locus defined by the equations

$$f_i(x) = 0 \qquad (i = 1, 2, \ldots, n).$$

On the other hand, any principal locus on V is the trace on V of a principal locus on $V^{(L)}$. The first assertion follows from these observations. The second assertion is clear because if ω vanishes at every point of V, then $\omega = 0$.

Corollary. *Let* $\omega_1, \omega_2 \in L[V^{(L)}]$. *Then* $\omega_1 = \omega_2$ *if and only if* $\omega_1(x) = \omega_2(x)$ *for all* $x \in V$.

The following lemma, which is too obvious to require proof, is a useful aid in identifying the results of extending the ground field.

Lemma 14. *Let* V *be an affine set defined over* K, V* *an affine set defined over* L, *and suppose that* V *is a subset of* V*. *Denote by* S *the* L-*algebra obtained by restricting the functions that make up* L[V*] *to* V *and suppose that the following two conditions are satisfied:*

(a) $S = K[V]^L$,

(b) *the homomorphism* $L[V^*] \to S$ *of* L-*algebras induced by restriction is an isomorphism.*

Then the obvious bijection between $V^{(L)}$ *and* V* *is an* L-*isomorphism.* *(Hence* $V^{(L)}$ *may be identified with* V*.)

We give an application of this lemma.

Theorem 30. *Let* U *be a closed subset of* V. *Then the closure of* U *in* $V^{(L)}$ *is* $U^{(L)}$.

Remark. It should be noted that if U is a principal locus on V it does not follow that $U^{(L)}$ will be a principal locus on $V^{(L)}$.

Proof. Let $\{\lambda_i\}_{i \in I}$ be a base for L over K and let $\omega \in L[V^{(L)}]$. Then ω has a unique representation in the form

$$\omega = \sum_{i \in I} \lambda_i f_i, \qquad (2.10.6)$$

where f_i is (the prolongation of) a member of K[V], and only finitely many f_i are non-zero. It follows that ω vanishes on U if and only if $f_i \in I_V(U)$ for all i. Consequently $\omega(U) = \{0\}$ precisely when $\omega \in I_V(U)L[V^{(L)}]$. Denote by U* the locus of the ideal $I_V(U)L[V^{(L)}]$ in $V^{(L)}$. Then Theorem 9 shows that U* is the closure of U in $V^{(L)}$.

Let S^* respectively S be the L-algebra that one obtains from $L[V^{(L)}]$ by restricting the domain of the functions to U^* respectively U. Then U^* is an affine set (defined over L) whose coordinate ring is S^* and we have a natural homomorphism $S^* \to S$ of L-algebras. This indeed is an isomorphism because if ω vanishes on U, then it must vanish on U^*.

Let $\omega \in L[V^{(L)}]$ and be represented as in (2.10.6). The restriction of ω to U is then

$$\sum_{i \in I} \lambda_i \overline{f}_i,$$

where $\overline{f}_i$ is the restriction of f_i to U. It follows that

$$S = K[U]^L$$

and now the theorem is a consequence of Lemma 14.

Corollary 1. *Let U be a closed subset of the affine set V and let $U^{(L)}$ be regarded as a closed subset of $V^{(L)}$. Then the ideal of $L[V^{(L)}]$ that is associated with $U^{(L)}$ (in the sense of Theorem 10) is $I_V(U)L[V^{(L)}]$.*

Proof. Let $\omega \in L[V^{(L)}]$. The proof just given shows that ω vanishes on $U^{(L)}$ if and only if it vanishes on U. It also shows that ω vanishes on U if and only if $\omega \in I_V(U)L[V^{(L)}]$.

Corollary 2. *Let U be a closed subset of the affine set V and let $U^{(L)}$ be regarded as a closed subset of $V^{(L)}$. Then $U^{(L)} \cap V = U$.*

Proof. From Theorems 29 and 30 we see that $U^{(L)} \cap V$ is the closure of U in V. Consequently $U^{(L)} \cap V = U$.

Corollary 3. *Let $U_1, U_2, \ldots, U_m$ be closed subsets of the affine set V and let L be an extension field of K. Then, as subsets of $V^{(L)}$, we have*

$$(U_1 \cup U_2 \cup \ldots \cup U_m)^{(L)} = U_1^{(L)} \cup U_2^{(L)} \cup \ldots \cup U_m^{(L)}.$$

We shall now examine the effect of enlarging K to L on K-morphisms. To this end let $\phi : V \to W$ be a K-morphism of affine sets and $\psi : V^{(L)} \to W^{(L)}$ an L-morphism of affine sets. We recall that ϕ induces a homomorphism $\phi^* : K[W] \to K[V]$ of K-algebras. Likewise ψ induces a homomorphism

$$\psi^* : L[W^{(L)}] \to L[V^{(L)}]$$

of L-algebras. As usual, if $x \in V$ and $y \in W$, then M_x and N_y will denote the corresponding rational maximal ideals in $K[V]$ and $K[W]$.

Lemma 15. Let the notation be as above. Then ψ extends $\phi : V \to W$ if and only if ψ^* extends $\phi^* : K[W] \to K[V]$.

Proof. Assume first that ψ^* extends ϕ^*. Let $x \in V$. By (2. 6. 3),

$$\phi^*(N_{\phi(x)}) \subseteq M_x$$

whence

$$\psi^*(N_{\phi(x)} L[W^{(L)}]) \subseteq M_x L[V^{(V)}]$$

and therefore $\psi(x) = \phi(x)$. Thus ψ extends ϕ.

Now assume that ψ extends ϕ and suppose that $\theta \in K[W] \subseteq L[W^{(L)}]$. Then $\phi^*(\theta) \in K[V] \subseteq L[V^{(L)}]$ and it will suffice to show that $\phi^*(\theta)$ and $\psi^*(\theta)$ coincide as members of $L[V^{(L)}]$. Let $x \in V$. Then, at x, $\phi^*(\theta)$ and $\psi^*(\theta)$ both take the value $\theta(\phi(x))$. That $\phi^*(\theta) = \psi^*(\theta)$ now follows from Theorem 29 Cor.

We know, from section (1. 6), that there is a unique homomorphism

$$K[W]^L \to K[V]^L$$

of L-algebras which extends ϕ^*. This yields the

Corollary. There is exactly one L-morphism $V^{(L)} \to W^{(L)}$ which extends a given K-morphism $\phi : V \to W$.

In view of this result we introduce the

Notation. The unique extension of a K-morphism $\phi : V \to W$ to an L-morphism $V^{(L)} \to W^{(L)}$ will be denoted by $\phi^{(L)}$.

Obviously if ϕ is an identity K-morphism, then $\phi^{(L)}$ is an identity L-morphism. Also if $\phi_1 : V \to V'$ and $\phi_2 : V' \to V''$ are K-morphisms, then

$$(\phi_2 \circ \phi_1)^{(L)} = \phi_2^{(L)} \circ \phi_1^{(L)}. \tag{2.10.7}$$

Lemma 16. Let $\phi : V \to W$ be a K-morphism. Then ϕ is a K-isomorphism if and only if $\phi^{(L)} : V^{(L)} \to W^{(L)}$ is an L-isomorphism.

Proof. Put $\psi = \phi^{(L)}$. It is trivial that when ϕ is a K-isomorphism ψ is an L-isomorphism. Assume that ψ is an L-isomorphism. Then, with the notation of Lemma 15, $\psi^* : L[W^{(L)}] \to L[V^{(L)}]$ is an isomorphism of L-algebras extending $\phi^* : K[W] \to K[V]$ and we have to show that ϕ^* is an isomorphism. Obviously ϕ^* is an injection.

Let $\{\lambda_i\}_{i \in I}$ be a base for L over K. Since the elements of K[W] span $L[W^{(L)}]$ as an L-space, and since ψ^* is surjective, we see that each $\omega \in L[V^{(L)}]$ can be expressed in the form

$$\omega = \sum_{i \in I} \lambda_i f_i$$

where $f_i \in \phi^*(K[W])$ and only finitely many of the f_i are non-zero. But $L[V^{(L)}] = K[V]^L$ and now it is clear that $\phi^*(K[W]) = K[V]$. This complet the proof.

Let us now consider the effect of the extension of the ground field on the product $V \times W$. Suppose therefore that $\{f_i\}_{i \in I}$ respectively $\{g_j\}_{j \in J}$ is a base for K[V] respectively K[W] over K; furthermore le F_i respectively G_j denote the prolongation of f_i respectively g_j to $V^{(L)}$ respectively $W^{(L)}$. Then $\{F_i\}_{i \in I}$ is a base for $L[V^{(L)}]$ over L and $\{G_j\}_{j \in J}$ a base for $L[W^{(L)}]$ over L. Accordingly the functions $F_i \vee G_j$ form a base for $L[V^{(L)} \times W^{(L)}]$ over L. On the other hand, th $f_i \vee g_j$ form a base for $K[V \times W]$ over K and therefore they form a base for $K[V \times W]^L$ over L.

Let S be the L-algebra formed by the restrictions of the members of $L[V^{(L)} \times W^{(L)}]$ to $V \times W$ and consider the homomorphism

$$L[V^{(L)} \times W^{(L)}] \rightarrow S \tag{2.10.8}$$

of L-algebras to which the process of restriction gives rise. Since $F_i \vee G_j$ maps into $f_i \vee g_j$ it follows that

$$S = K[V \times W]^L.$$

Also, because a base of $L[V^{(L)} \times W^{(L)}]$ over L is mapped into a base of S over L, (2.10.8) is an isomorphism. We may therefore conclude, by appealing to Lemma 14, that

$$V^{(L)} \times W^{(L)} = (V \times W)^{(L)}. \tag{2.10.9}$$

We illustrate the consequences of enlarging the ground field by considering two special cases.

Theorem 31. <u>If the affine set</u> V <u>is finite, then</u> $V^{(L)} = V$.

Proof. By Theorem 18, we have $K[V]^L = \mathfrak{F}_L(V)$. However V can be regarded as an affine set defined over L with $\mathfrak{F}_L(V)$ as its coordinate ring. Thus, in this case, making the ground field bigger produces no new points.

For our second example, we let V be an n-dimensional vector space over K. Then, as we saw in section (2.5), V has a natural structure as an affine set. Let $\xi_1, \xi_2, \ldots, \xi_n$ be a base for $\mathrm{Hom}_K(V, K)$ over K. Then, by Theorem 19, $K[V] = K[\xi_1, \xi_2, \ldots, \xi_n]$. Consequently $K[V]^L = L[\xi_1, \xi_2, \ldots, \xi_n]$.

Define V^L as in section (1.6.1). This is an n-dimensional vector space over L, each $\xi_i : V \rightarrow K$ has a unique extension to an L-linear mapping $\hat{\xi}_i : V^L \rightarrow L$, and $\hat{\xi}_1, \hat{\xi}_2, \ldots, \hat{\xi}_n$ is a base for $\mathrm{Hom}_L(V^L, L)$. Accordingly, by Theorem 19,

$$L[V^L] = L[\hat{\xi}_1, \hat{\xi}_2, \ldots, \hat{\xi}_n].$$

Let S be the L-algebra obtained by restricting the domain of the functions forming $L[V^L]$ to V. The natural surjective homomorphism

$$L[V^L] = L[\hat{\xi}_1, \hat{\xi}_2, \ldots, \hat{\xi}_n] \rightarrow S$$

of L-algebras which results is such that $\hat{\xi}_i \mapsto \xi_i$. Consequently

$$S = L[\xi_1, \xi_2, \ldots, \xi_n] = K[V]^L.$$

Theorem 32. *Let V be an n-dimensional vector space over K. Then V can be regarded as an affine set defined over K and V^L as an affine set defined over L. If the field K is infinite, then $V^L = V^{(L)}$.*

Remark. Theorem 31 shows that the requirement that K be infinite cannot be left out.

Proof. We use the same notation as that employed in the introduction to the theorem, and note at once that, because of Lemma 14, it will suffice to prove that $\xi_1, \xi_2, \ldots, \xi_n$ are algebraically independent over L. But, by Theorem 19, the power products $\xi_1^{\nu_1}\xi_2^{\nu_2}\ldots\xi_n^{\nu_n}$ are linearly independent over K and so, when considered as belonging to $K[V]^L$, they will be linearly independent over L. The theorem follows.

We add a few general comments. Let $\phi : V \to W$ be a K-morphism of affine sets. The closure $\overline{\phi(V)}$, of $\phi(V)$ in W, is an affine set and $\phi^{(L)} : V^{(L)} \to W^{(L)}$ is an L-morphism. Because ϕ can be factored through $\overline{\phi(V)}$, it follows that

$$\phi^{(L)}(V^{(L)}) \subseteq \overline{\phi(V)}^{(L)}$$

and therefore, with a self-explanatory notation,

$$\overline{\phi^{(L)}(V^{(L)})} \subseteq \overline{\phi(V)}^{(L)}.$$

On the other hand $\phi(V) \subseteq \phi^{(L)}(V^{(L)})$ and hence

$$\overline{\phi(V)} \subseteq \overline{\phi^{(L)}(V^{(L)})}.$$

Consequently, by Theorem 30,

$$\overline{\phi(V)}^{(L)} \subseteq \overline{\phi^{(L)}(V^{(L)})}$$

and thus we have

$$\overline{\phi(V)}^{(L)} = \overline{\phi^{(L)}(V^{(L)})}. \tag{2.10.10}$$

One final observation. In addition to assuming that L is an extension field of K, let L' be an extension field of L. Then

$$(V^{(L)})^{(L')} = V^{(L')} \tag{2.10.11}$$

This follows from the relation

$$(K[V]^L)^{L'} = K[V]^{L'}.$$

Of course $V^{(K)} = V$.

2.11 Generalized points and generic points

Let V be an affine set defined over K and let L be an extension field of K. A point ξ of $V^{(L)}$ is called a <u>generalized point</u> of V. Suppose that ξ is such a point. Then there is a homomorphism

$$K[V] \to L \tag{2.11.1}$$

of K-algebras in which $f \in K[V]$ is mapped into $f(\xi)$. The image of K[V] under (2.11.1) is denoted by $K[\xi]$. We use $K(\xi)$ to denote the quotient field of $K[\xi]$.

Conversely let

$$\sigma : K[V] \to L \tag{2.11.2}$$

be a homomorphism of K-algebras. Then there exists a homomorphism $K[V] \otimes_K L \to L$, of L-algebras, in which $f \otimes \lambda \mapsto \lambda\sigma(f)$; that is we have a homomorphism

$$L[V^{(L)}] \to L, \tag{2.11.3}$$

of L-algebras, which on K[V] reduces to σ. The kernel of (2.11.3) corresponds to a point $\xi \in V^{(L)}$ and σ itself is the mapping $K[V] \to L$ in which $f \mapsto f(\xi)$.

Definition. Let $\xi \in V^{(L)}$. We say that ξ is a 'generic point' of V if the homomorphism (2.11.1) is an injection.

Thus a generalized point ξ, of V, is a generic point if and only if it has the following property: whenever $f \in K[V]$ and $f(\xi) = 0$, then $f = 0$. Evidently if ξ is a generic point of V, then K[V] and $K[\xi]$ are isomorphic K-algebras. In particular K[V] is an integral domain.

Theorem 33. If V has a generic point, then K[V] is an integral domain. If K[V] is an integral domain and L_0 is an extension field of K, then there exists an extension field L, of L_0, such that $V^{(L)}$ contains a generic point of V.

Proof. We need only establish the final assertion. Let K[V] be an integral domain. We can certainly find an extension field L, of L_0, and a homomorphism $\sigma : K[V] \to L$ of K-algebras which is an injection. As we saw above, there exists $\xi \in V^{(L)}$ such that $\sigma(f) = f(\xi)$ for all $f \in K[V]$. Thus ξ is a generic point of V.

Now suppose that each of L and Ω is an extension field of K. Let $\xi \in V^{(L)}$, $\eta \in V^{(\Omega)}$ so that ξ, η are generalized points of V.

Definition. We say that 'η is a specialization of ξ' if whenever $f \in K[V]$ and $f(\xi) = 0$, then $f(\eta) = 0$.

Accordingly η is a specialization of ξ if and only if there is a homomorphism $K[\xi] \to [\eta]$, of K-algebras, which makes

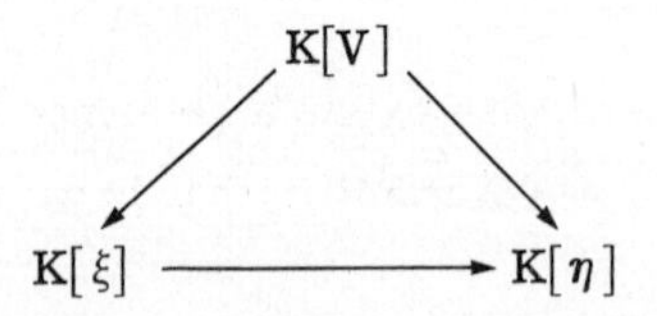

a commutative diagram. It follows that if ξ is a generic point of V, then every point of V (generalized or ordinary) is a specialization of ξ.

Suppose now that ξ, η, ζ are generalized points of V, that η is a specialization of ξ, and ζ is a specialization of η. Then ζ is a specialization of ξ. Again, if each of ξ, η is a specialization of the other, then the homomorphism $K[\xi] \to K[\eta]$ of K-algebras is an isomorphism and $K[\eta] \to K[\xi]$ is its inverse. Hence if η is a specialization of ξ and η is a generic point of V, then ξ is also a generic point of V.

Theorem 34. *Let V be an affine set defined over K, let L be an extension field of K, and let $\xi \in V^{(L)}$. If now U is a closed subset of V, then the following two statements are equivalent:*

(a) $\xi \in U^{(L)}$;

(b) $f(\xi) = 0$ *whenever* $f \in I_V(U)$.

Proof. Assume (a) and let $f \in I_V(U)$. Then f, considered as a member of $L[V^{(L)}]$, vanishes on the closure of U in $V^{(L)}$, that is to say it vanishes on $U^{(L)}$. In particular $f(\xi) = 0$.

Next suppose that (b) holds. Then ξ is a common zero of the members of $I_V(U)L[V^{(L)}]$. But, by Theorem 30 Cor. 1, $I_V(U)L[V^{(L)}]$ is the ideal of $L[V^{(L)}]$ that is associated with $U^{(L)}$ and therefore its zeros are the points of $U^{(L)}$. It follows that ξ belongs to $U^{(L)}$ and so the proof is complete.

Corollary 1. *Let ξ, ξ' belong to $V^{(L)}$, $V^{(\Omega)}$ respectively*† *and suppose that ξ' is a specialization of ξ when these are considered as generalized points of V. Let U be a closed subset of V. If now $\xi \in U^{(L)}$, then $\xi' \in U^{(\Omega)}$ and ξ' is a specialization of ξ when they are considered as generalized points of U.*

Proof. Suppose that $f \in K[V]$ and let g be its contraction to U. Then $f(\xi) = g(\xi)$. The corollary follows from this observation.

Corollary 2. *Let U be a closed subset of V and ξ a generalized point of U and hence also of V. Then the specializations of ξ when it is considered as a generalized point of V are the same as its specializations when considered as a generalized point of U.*

Corollary 3. *Let U be a closed subset of V and ξ a generalized point of U and hence also of V. Then $K[\xi]$ is the same whether ξ is regarded as belonging to U or as belonging to V.*

---

† It is understood that Ω, as well as L, is an extension field of K.

Suppose next that V, W are affine sets defined over K. (As before L denotes an extension field of K.) Let $\xi, \xi' \in V^{(L)}$ and $\eta, \eta' \in W^{(L)}$. By (2.10.9), (ξ, η) and (ξ', η') belong to $(V \times W)^{(L)}$. Evidently the K-algebra $K[(\xi, \eta)]$ contains $K[\xi]$ and $K[\eta]$ as subalgebras and the smallest subring of $K[(\xi, \eta)]$ that contains both $K[\xi]$ and $K[\eta]$ is $K[(\xi, \eta)]$ itself. Also if (ξ', η') is a specialization of (ξ, η), then ξ' is a specialization of ξ and η' is a specialization of η.

Consider a K-morphism $\phi : V \to W$ and let $\xi \in V^{(L)}$, where L is an extension field of K. By section (2.10), ϕ determines an L-morphism $\phi^{(L)} : V^{(L)} \to W^{(L)}$ which extends ϕ. The point $\phi^{(L)}(\xi)$ is usually denoted by $\phi(\xi)$. Thus $\phi(\xi)$ is a generalized point of W and if $g \in K[W]$, then

$$g(\phi(\xi)) = f(\xi), \qquad (2.11.4)$$

where $f = g \circ \phi \in K[V]$.

Lemma 17. _Let_ $\phi : V \to W$ _be a_ K-_morphism of affine sets and_ ξ _a generalized point of_ V. _Then_ $K[\phi(\xi)] \subseteq K[\xi]$.

Proof. Let $g \in K[W]$. Then $g(\phi(\xi)) \in K[\xi]$ by (2.11.4) and the lemma follows.

In the next theorem, L and Ω denote extension fields of K.

Theorem 35. _Let_ $\phi : V \to W$ _be a_ K-_morphism of affine sets and suppose that_ $\xi \in V^{(L)}$, $\eta \in V^{(\Omega)}$. _If_ η _is a specialization of_ ξ, _then_ $\phi(\eta)$ _is a specialization of_ $\phi(\xi)$.

Proof. Let $g \in K[W]$ and put $f = g \circ \phi$ so that $f \in K[V]$. Assume that $g(\phi(\xi)) = 0$. By (2.11.4) we have $f(\xi) = 0$ and therefore $f(\eta) = 0$. But then, again by (2.11.4), $g(\phi(\eta)) = 0$ and the theorem is proved.

Sometimes after the ground field has been enlarged to achieve one particular end, we later wish to contract it partially in order to achieve another. In this connection the following lemma can be useful.

Lemma 18. _Let_ V _be an affine set defined over_ K, _let_ L _be an extension field of_ K, _and_ L' _and extension field of_ L. _If now_ $\xi' \in V^{(L')}$

and $K[\xi'] \subseteq L$, then $\xi' \in V^{(L)}$.

Proof. Let $\omega : L'[V^{(L')}] \to L'$ be the homomorphism of L'-algebras which maps each function into its value at ξ'. For $f \in K[V]$, we have $f(\xi') \in K[\xi'] \subseteq L$. It follows that restriction leads to a homomorphism $L[V^{(L)}] \to L$ of L-algebras and so there exists $\xi \in V^{(L)}$ such that $f(\xi) = f(\xi')$ for all $f \in K[V]$. But then $\theta(\xi) = \theta(\xi')$ for all θ in $L'[V^{(L')}]$ and hence $\xi = \xi'$.

3 · Irreducible affine sets

General remarks

The next stage in the development of the theory of affine sets will be the application of the theory of irreducible topological spaces. Since the properties of such spaces may be unfamiliar to the reader, all the results that we shall need will be established in a short preliminary section.

Once again K denotes a field concerning which no special assumptions are made unless these are stated explicitly in connection with particular situations.

3.1 Irreducible spaces

Throughout section (3.1) X denotes a topological space. If U is a subset of X, then $\overline{U}$ denotes its closure.

Definition. The space X is called 'irreducible' if (i) it is not empty, and (ii) whenever X is expressed in the form $X = F_1 \cup F_2$, where F_1, F_2 are closed subsets, either $F_1 = X$ or $F_2 = X$.

Theorem 1. *An irreducible topological space is connected. Also any continuous image of an irreducible space is itself irreducible.*

These assertions follow immediately from the definition.

Theorem 2. *Let X be irreducible and let O be a non-empty open subset. Then $X = \overline{O}$.*

This is clear because X is the union of the closed sets $\overline{O}$ and $X \setminus O$.

Theorem 3. *Let* X *be irreducible and let* O_1, O_2 *be non-empty open subsets. Then* $O_1 \cap O_2$ *is non-empty.*

Proof. By Theorem 2, $\overline{O}_2 = X$. Let $x \in O_1$. Then x belongs to the closure of O_2 and O_1 is an open neighbourhood of x. Consequently O_1 and O_2 intersect.

Let Y be a subset of X. Then Y can be regarded as a subspace.

Definition. We say that Y is an 'irreducible subset' of X if, when regarded as a subspace, Y is an irreducible topological space.

For example, if $x \in X$, then $\{x\}$ is an irreducible subset of X.

Lemma 1. *Let* Y *be a non-empty subset of* X. *Then the following statements are equivalent:*

(a) Y *is an irreducible subset of* X;

(b) *if* $Y \subseteq F_1 \cup F_2$, *where* F_1, F_2 *are closed in* X, *then either* $Y \subseteq F_1$ *or* $Y \subseteq F_2$.

This is no more than a restatement of the definition of an irreducible subset.

Lemma 2. *Let* Y *be an irreducible subset of* X *and suppose that* $Y \subseteq F_1 \cup F_2 \cup \ldots \cup F_m$, *where* $F_1, F_2, \ldots, F_m$ *are closed subsets of* X. *Then* $Y \subseteq F_i$ *for some* i.

Lemma 2 follows at once from Lemma 1.

Theorem 4. *Let* Y *be an irreducible subset of* X. *Then* $\overline{Y}$ *is also an irreducible subset of* X.

This also follows from Lemma 1.

Lemma 3. *Let* $\{Y_i\}_{i \in I}$ *be a family of irreducible subsets of* X, *where* (i) I *is not empty, and* (ii) $Y_{i_1} \subseteq Y_{i_2}$ *or* $Y_{i_2} \subseteq Y_{i_1}$ *whenever* $i_1, i_2 \in I$. *Put*

$$Y = \bigcup_{i \in I} Y_i.$$

Then Y *is an irreducible subset of* X.

Proof. Suppose that $Y \subseteq F_1 \cup F_2$, where F_1, F_2 are closed subsets of X, and assume that $Y \not\subseteq F_1$ and $Y \not\subseteq F_2$. We must derive a contradiction.

There exist $i_1, i_2 \in I$ such that $Y_{i_1} \not\subseteq F_1$ and $Y_{i_2} \not\subseteq F_2$. Now either $Y_{i_1} \subseteq Y_{i_2}$ or $Y_{i_2} \subseteq Y_{i_1}$. We shall suppose, for definiteness, tha $Y_{i_1} \subseteq Y_{i_2}$. Since $Y_{i_2} \subseteq F_1 \cup F_2$, Y_{i_2} is irreducible, and $Y_{i_2} \not\subseteq F_2$, it follows that $Y_{i_2} \subseteq F_1$. But then $Y_{i_1} \subseteq F_1$ and we have obtained the desired contradiction.

Lemma 3 shows that the irreducible subsets of X, when partially ordered by inclusion, form an inductive system.

Definition. The maximal irreducible subsets of X are called the 'irreducible components' of X.

Theorem 5. Each irreducible subset of X is contained in an irreducible component of X. The irreducible components of X are closed subsets, and X itself is their union.

Proof. Let Y be an irreducible subset of X. The irreducible subsets of X that contain Y form a non-empty inductive system and each maximal member of this inductive system will be an irreducible component of X. This proves the first assertion. The second assertion follows from Theorem 4. Finally if $x \in X$, then, by the first part, there is an irreducible component of X containing $\{x\}$. This completes the proof.

Let us suppose that we have a representation of X in the form

$$X = X_1 \cup X_2 \cup \ldots \cup X_m, \tag{3.1.1}$$

where each X_i is a closed irreducible subset of X. Then the irreducibl components of X are the maximal members of the set $\{X_1, X_2, \ldots, X$ and, in particular, the number of irreducible components in this case is finite. Moreover if the representation (3.1.1) is irredundant, so that for each value of i

$$X \neq X_1 \cup \ldots \cup X_{i-1} \cup X_{i+1} \cup \ldots \cup X_m,$$

then $X_1, X_2, \ldots, X_m$ are precisely the irreducible components of X. Conversely if X has only a finite number of irreducible components, then not only is X their union but also the union is irredundant.

We now introduce a condition on X that ensures that the number of its irreducible components is finite.

Definition. The space X is said to be 'Noetherian' if it satisfies the minimal condition for closed subsets.

Thus if X is Noetherian, then every non-empty collection of closed subsets contains at least one minimal member. Further any subspace formed by a closed subset will also be Noetherian.

Theorem 6. Let X be a Noetherian topological space. Then X has only a finite number of irreducible components.

Proof. Let Y be a closed subset of X. It will be shown that Y is the union of a finite number of closed irreducible subsets of X. This will establish the theorem.

Assume the contrary. Then among the closed subsets Y which do not have a representation of the right kind there will exist one, Y_0 say, which is minimal with respect to inclusion. Now Y_0 is not empty nor can it be irreducible. Hence $Y_0 = F_1 \cup F_2$, where $F_1 \neq Y_0$, $F_2 \neq Y_0$ and each of F_1, F_2 is closed in X. But, by the minimality of Y_0, F_1, and likewise F_2, is a finite union of closed irreducible subsets of X. But this implies that $F_1 \cup F_2 = Y_0$ is also a finite union of closed irreducible subsets and now we have the contradiction that we were seeking.

3.2 Irreducible affine sets

Throughout section (3.2) V will denote an affine set defined over K. Since V is a topological space by virtue of its affine topology, the results of section (3.1) may be applied to it. First, however, it will be convenient to recall some basic results from the theory of commutative rings.

To this end let R be a commutative ring with an identity element.

It is known[†] that the following statements concerning R are equivalent:

(a) every ideal is finitely generated;

(b) every infinite increasing sequence of ideals becomes constant

(c) every non-empty set of ideals contains at least one maximal member.

Should it happen that R possesses these properties, then it is said to be a Noetherian ring. Clearly any homomorphic image of a Noetherian ring is itself Noetherian.

Next suppose that $X_1, X_2, \ldots, X_n$ are indeterminates. The celebrated Basis Theorem[‡] states that if R is a Noetherian ring, then the polynomial ring $R[X_1, X_2, \ldots, X_n]$ is also Noetherian. Since the field K is trivially Noetherian, it follows that $K[X_1, X_2, \ldots, X_n]$ is Noetherian. But Chapter 2 Lemma 4 shows that K[V] is a homomorphic image of such a ring. Consequently we have proved

Theorem 7. The coordinate ring of an affine set V is a Noetherian ring.

Next, by Chapter 2 Theorem 10, there is a bijection between the set of closed subsets of V on the one hand and the set formed by certain ideals of K[V] on the other; furthermore the bijection described in the theorem reverses inclusion relations. Hence, in view of Theorem 7, we obtain

Theorem 8. With respect to its affine topology, the affine set V is a Noetherian space.

It now follows, from Theorem 6, that V has only a finite number of irreducible components. We say that V is an irreducible affine set if it has precisely one irreducible component. In other words V is an irreducible affine set if and only if it is irreducible as a topological space.

---

† See, for example, [(9) Proposition 9, p. 22].

‡ For a proof see [(9) Theorem 1, p. 179].

Theorem 9. <u>The following statements, concerning the affine set V, are equivalent:</u>

(i) V <u>is irreducible;</u>

(ii) $K[V]$ <u>is an integral domain;</u>

(iii) V <u>possesses a generic point.</u>

Proof. We already know, from Chapter 2 Theorem 33, that (ii) and (iii) are equivalent. In what follows we demonstrate the equivalence of (i) and (ii).

<u>First assume (i).</u> Suppose that $f, g \in K[V]$ and $fg = 0$. If we put

$$F_1 = \{x \mid x \in V \text{ and } f(x) = 0\},$$

$$F_2 = \{x \mid x \in V \text{ and } g(x) = 0\},$$

then F_1, F_2 are closed subsets of V and $F_1 \cup F_2 = V$. Thus either $F_1 = V$ or $F_2 = V$, that is to say either $f = 0$ or $g = 0$. Accordingly (i) implies (ii).

<u>Now assume (ii).</u> We suppose that $V = F_1 \cup F_2$, where F_1, F_2 are closed subsets of V. It will suffice to prove that $F_1 = V$ or $F_2 = V$. We assume the contrary and seek a contradiction.

Since $F_1 \neq V$, it follows (see Chapter 2 Theorem 10) that $I_V(F_1) \neq (0)$. Likewise $I_V(F_2) \neq (0)$. On the other hand

$$I_V(F_1) \cap I_V(F_2) = I_V(V) = (0)$$

and therefore $I_V(F_1)I_V(F_2) = (0)$. But this is contrary to our assumption that $K[V]$ is an integral domain.

Corollary 1. <u>Let</u> U <u>be a closed subset of the affine set</u> V. <u>Then</u> U <u>is irreducible if and only if</u> $I_V(U)$ <u>is a prime ideal.</u>

This follows from Theorem 9 because $K[U]$ is isomorphic to $K[V]/I_V(U)$.

Corollary 2. <u>Let</u> $\phi : V \to W$ <u>be a</u> K-<u>morphism of affine sets, where</u> V <u>is irreducible. Then the closure</u> $\overline{\phi(V)}$, <u>of</u> $\phi(V)$ <u>in</u> W, <u>is irreducible Moreover if</u> ξ <u>is a generic point of</u> V, <u>then</u> $\phi(\xi)$ <u>is a</u>

generic point of $\overline{\phi(V)}$.

Proof. The first assertion follows from Theorems 1 and 4. Next, by (2.10.10), $\phi(\xi)$ is a generalized point of $\overline{\phi(V)}$. Put

$$\mathfrak{A} = \{f \mid f \in K[W] \text{ and } f(\phi(\xi)) = 0\}$$

and let $x \in V$. Then, by Chapter 2 Theorem 35, $\phi(x)$ is a specialization of $\phi(\xi)$ and therefore $\phi(x) \in C_W(\mathfrak{A})$. It follows that $\overline{\phi(V)} \subseteq C_W(\mathfrak{A})$.

Finally let $g \in K[\overline{\phi(V)}]$ be the restriction, to $\overline{\phi(V)}$, of $f \in K[W]$ and suppose that $g(\phi(\xi)) = 0$. Then $f \in \mathfrak{A}$ and hence it vanishes on $\overline{\phi(V)}$. Thus $g = 0$ and the proof is complete.

At this point we interrupt the development of the main theory to mention two examples.

Theorem 10. *If* V *is a finite affine set, then its irreducible components are its individual points.*

This is obvious.

Theorem 11. *Let* V *be a finite-dimensional vector space over* K. *If* K *is an infinite field and* V *is considered as an affine set, then* V *is irreducible.*

Proof. Let $X_1, X_2, \ldots, X_n$ be indeterminates, where n is the dimension of V as a vector space. Then $K[V]$ and $K[X_1, X_2, \ldots, X_n]$ are isomorphic K-algebras. Consequently $K[V]$ is an integral domain.

We now return to the general theory.

Theorem 12. *There is a natural bijection between the closed irreducible subsets of* V *and the prime ideals* P, *of* $K[V]$, *which satisfy* $\mathrm{Rad}_V(P) = P$. *If the closed irreducible subset* U *is associated with the prime ideal* P *(in this bijection), then* $U = C_V(P)$ *and* $P = I_V(U)$.

This follows by combining Theorem 9 Cor. 1 with Theorem 10 of Chapter 2.

Corollary. *If* K *is algebraically closed, the bijection described in Theorem 12 is between the set of closed irreducible subsets of* V *and* _the set of all prime ideals of_ K[V].

Proof. Since K is algebraically closed, every prime ideal P, of K[V], satisfies† $Rad_V(P) = P$.

Let U be a closed subset of V. From Theorems 6 and 8 we see that U has only a finite number of irreducible components. Denote these by $U_1, U_2, \ldots, U_m$. Then

$$U = U_1 \cup U_2 \cup \ldots \cup U_m. \tag{3.2.1}$$

Put

$$P_i = I_V(U_i). \tag{3.2.2}$$

Since U_i is an irreducible closed subset of V, P_i is a prime ideal of K[V] and $Rad_V(P_i) = P_i$. Also if $i \neq j$, then $P_i \not\subseteq P_j$.

From (3.2.1) we conclude that

$$I_V(U_1) \cap I_V(U_2) \cap \ldots \cap I_V(U_m) = I_V(U)$$

that is to say

$$P_1 \cap P_2 \cap \ldots \cap P_m = I_V(U). \tag{3.2.3}$$

Let P be any prime ideal of K[V] that contains $I_V(U)$. Then $P_1 P_2 \ldots P_m \subseteq P$ and therefore $P_i \subseteq P$ for some value of i. Thus $P_1, P_2, \ldots, P_m$ are the minimal members of the set of all prime ideals that contain $I_V(U)$, that is to say they are the _minimal prime ideals_ of $I_V(U)$. These observations add up to a proof of

Theorem 13. _Let_ U _be a closed subset of_ V. _Then_ $I_V(U)$ _has only a finite number of minimal prime ideals. Let these be_ $P_1, P_2, \ldots, P_m$. _Then_ $Rad_V(P_i) = P_i$ _for each value of_ i, _the irreducible affine sets_

$$C_V(P_1),\ C_V(P_2),\ \ldots,\ C_V(P_m)$$

are the irreducible components of U, _and_

---

† See the remarks following Theorem 13 of Chapter 2.

$$I_V(U) = P_1 \cap P_2 \cap \dots \cap P_m.$$

The case $U = V$ deserves a special mention.

Corollary. The irreducible components of V correspond to the minimal prime ideals of the zero ideal of $K[V]$.

Theorem 14. Let V be an irreducible affine set, let $f \in K[V]$ and suppose that f vanishes on some non-empty open subset N of V. Then $f = 0$.

Proof. The principal locus

$$\{x \mid x \in V \text{ and } f(x) = 0\}$$

is closed and it contains N. It therefore contains the closure $\overline{N}$ of N. However, by Theorem 2, $\overline{N} = V$.

Theorem 15. Let $x \in V$ and $f \in K[V]$. Suppose that f vanishes on some open set N, where $x \in N$. Then there exists $g \in K[V]$ such that $g(x) \neq 0$ and $fg = 0$.

Proof. Let $V_1, V_2, \dots, V_s$ be the irreducible components of V and put $P_i = I_V(V_i)$. Further let M_x be the rational maximal ideal of $K[V]$ that corresponds to x. Without loss of generality we may suppose that

$$P_1 \not\subseteq M_x,\ P_2 \not\subseteq M_x,\ \dots,\ P_\nu \not\subseteq M_x$$

and

$$P_{\nu+1} \subseteq M_x,\ P_{\nu+2} \subseteq M_x,\ \dots,\ P_s \subseteq M_x.$$

Suppose that $\nu+1 \leq j \leq s$. Then $x \in V_j$ and f vanishes on $V_j \cap N$. Consequently, by Theorem 14, f vanishes on the whole of V_j and therefore $f \in P_j$. Choose g in $P_1 \cap P_2 \cap \dots \cap P_\nu$ so that $g \notin M_x$. Thus $g(x) \neq 0$ and fg belongs to

$$P_1 \cap P_2 \cap \dots \cap P_s = I_V(V) = (0),$$

i.e. $fg = 0$. This completes the proof.

The next theorem provides a good example of the power of topological methods. In it V and W denote affine sets defined over K. Let $x \in V$ and denote by

$$\tau_x : W \to V \times W$$

the mapping in which $\tau_x(y) = (x, y)$. We know, from Chapter 2 Lemma 12, that τ_x is a K-morphism.

Theorem 16. _Let V and W be irreducible affine sets. Then $V \times W$ is also an irreducible affine set._

Proof. We must show that, when $V \times W$ is endowed with the affine topology, it is an irreducible space. Suppose therefore that $V \times W = T \cup T'$, where T and T' are closed subsets of $V \times W$. It will suffice to show that either $V \times W \subseteq T$ or $V \times W \subseteq T'$.

Let $y \in W$. By Chapter 2 Theorem 28, V and $V \times \{y\}$ are K-isomorphic and therefore they are homeomorphic topological spaces. Consequently $V \times \{y\}$ is irreducible and now it follows that either $V \times \{y\} \subseteq T$ or $V \times \{y\} \subseteq T'$.

Define subsets W_1 and W_2, of W, by

$$W_1 = \{y \mid y \in W \text{ and } V \times \{y\} \subseteq T\}$$

and

$$W_2 = \{y \mid y \in W \text{ and } V \times \{y\} \subseteq T'\}.$$

Then $W = W_1 \cup W_2$, by virtue of the observations made in the last paragraph, and it will now suffice to show that either $W_1 = W$ or $W_2 = W$. Moreover this will follow if we show that W_1 and W_2 are closed in W. But

$$W_1 = \bigcap_{x \in V} \tau_x^{-1}(T)$$

and this is closed because each τ_x is continuous. For similar reasons W_2 is closed. The theorem follows.

Corollary. *Let* R *and* S *be affine algebras over* K *and suppose that* R *and* S *are integral domains. Then* $R \otimes_K S$ *is an integral domai*

As the next result shows, Theorem 16 is easily generalized.

Theorem 17. *Let* $V_1, V_2, \ldots, V_r$ *be the irreducible components of the affine set* V *and* $W_1, W_2, \ldots, W_s$ *the irreducible components of the affine set* W. *Then the sets* $V_i \times W_j$, *where* $1 \leq i \leq r$, $1 \leq j \leq s$, *are the irreducible components of* $V \times W$.

Proof. The $V_i \times W_j$ are closed subsets of $V \times W$ and, by Theorem 16, they are irreducible. Furthermore $V \times W$ is their union. Since none of $V_1 \times W_1, V_1 \times W_2, \ldots, V_r \times W_s$ is contained by any of the others, the theorem follows.

3.3 Localization

Let V be an affine set defined over K and let $x \in V$. Suppose that $f, g \in K[V]$ and $g(x) \neq 0$. Then

$$N = \{y \mid y \in V \text{ and } g(y) \neq 0\}$$

is an open subset of V and $x \in N$. Furthermore we can define a mappin

$$\theta : N \to K \tag{3.3.1}$$

by

$$\theta(y) = \frac{f(y)}{g(y)} \,. \tag{3.3.2}$$

Now keep V and x fixed and let f and g vary. In this way we obtain a collection, $\Lambda_{V,x}$ say, of K-valued functions. Suppose that θ, θ' are in $\Lambda_{V,x}$. We introduce an equivalence relation on $\Lambda_{V,x}$ by regarding θ and θ' as equivalent if they agree on some non-empty open subset of V that contains x. Denote the set of equivalence classes by $Q_{V,x}$ and let $\Theta \in Q_{V,x}$. Then the functions which make up Θ all take the same value at x. If this common value is denoted by $\Theta(x)$, then we have a mapping

$$Q_{V,x} \to K \tag{3.3.3}$$

in which $\Theta \mapsto \Theta(x)$.

Next assume that f_1, g_1, f_2, g_2 all belong to $K[V]$ and that $g_1(x) \neq 0$, $g_2(x) \neq 0$. The pair f_i, g_i determines a function, θ_i say, that belongs to $\Lambda_{V,x}$. If N denotes the open set where $g_1 g_2$ does not vanish, then $x \in N$ and both θ_1 and θ_2 are defined on N. Furthermore $\theta_1 + \theta_2$ and $\theta_1 \theta_2$, considered as K-valued functions defined on N, belong to $\Lambda_{V,x}$.

It will now be clear how addition and multiplication may be defined on $Q_{V,x}$ and it is also clear that $Q_{V,x}$ has a natural structure as a K-space. In fact with this additional structure $Q_{V,x}$ is a K-algebra† and (3.3.3) is a surjective homomorphism of K-algebras.

Definition. The K-algebra $Q_{V,x}$ is called the 'local ring' of V at x.

If $f \in K[V]$, then f and 1 together determine a function (defined on V) which belongs to $\Lambda_{V,x}$. In this way we can associate with each $f \in K[V]$ an element of $Q_{V,x}$ and thus arrive at a homomorphism

$$K[V] \to Q_{V,x} \tag{3.3.4}$$

of K-algebras. It is customary to refer to (3.3.4) as the _canonical homomorphism_ of $K[V]$ into $Q_{V,x}$.

The above approach to localization is based on functions. We shall now give an alternative procedure that is more algebraic. First, however, we need

Theorem 18. _Let_ f_1, g_1, f_2, g_2 _belong to_ $K[V]$ _and assume that_ $g_1(x) \neq 0$, $g_2(x) \neq 0$. _Further let the pair_ f_i, g_i _determine the function_ θ_i _in_ $\Lambda_{V,x}$. _Then each of the following statements implies the other:_

(i) θ_1 _and_ θ_2 _are equivalent members of_ $\Lambda_{V,x}$;

(ii) _there exists_ $h \in K[V]$ _such that_ $h(x) \neq 0$ _and_ $hg_2 f_1 = hg_1 f_2$.

† This algebra is unitary, associative and commutative.

Proof. If (ii) holds, then θ_1 and θ_2 agree on the open set where hg_1g_2 does not vanish. Thus (ii) implies (i).

Now suppose that (i) holds. Then $f_1g_2 - f_2g_1$ vanishes on some open set containing x. That there exists $h \in K[V]$ with the required properties is therefore a consequence of Theorem 15. This completes the proof.

At this point we make a fresh start. Let $x \in V$ and denote by M_x the corresponding K-rational maximal ideal of K[V]. We now consider formal fractions f/g, where $f, g \in K[V]$ and $g \notin M_x$. These are just ordered pairs (written in a certain way) of elements of K[V], where the second component (the denominator) satisfies $g(x) \neq 0$. Two such formal fractions, say f/g and f'/g', are to be considered as equivalent if there exists $h \in K[V]$, such that (i) $h \notin M_x$, and (ii) $hg'f = hgf'$. (This is an equivalence relation in the abstract sense.) The set of equivalence classes will be denoted by $(K[V])_{M_x}$ and we use f/g not only to denote the formal fraction determined by f and g, but also the equivalence class to which the fraction belongs.

Suppose that f_1/g_1 and f_2/g_2 belong to $(K[V])_{M_x}$ and that $k \in K$. It is easily verified that $(K[V])_{M_x}$ constitutes an associative, unitary, and commutative K-algebra in which

$$\frac{f_1}{g_1} + \frac{f_2}{g_2} = \frac{g_2f_1 + g_1f_2}{g_1g_2},$$

$$\frac{f_1}{g_1}\,\frac{f_2}{g_2} = \frac{f_1f_2}{g_1g_2},$$

$$k\left(\frac{f_1}{g_1}\right) = \frac{(kf_1)}{g_1}.$$

Next let f/g be a formal fraction and θ the member of $\Lambda_{V,x}$ which it determines. Then, by Theorem 18, fractions equivalent to f/g determine functions equivalent to θ and the outcome is that we obtain an isomorphism

$$(K[V])_{M_x} \approx Q_{V,x} \tag{3.3.5}$$

of K-algebras. This isomorphism is frequently used to identify the two sides of (3.3.5). If this is done, then the homomorphism (3.3.3) maps f/g into $f(x)/g(x)$ whereas (3.3.4) takes f, of $K[V]$, into $f/1$.

Suppose that $f/g \in Q_{V,x}$, where $f, g \in K[V]$ and $g(x) \neq 0$. Obviously f/g is a unit in $Q_{V,x}$ if and only if $f(x) \neq 0$. Consequently f/g is a non-unit if and only if $f \in M_x$. Denote by $M_x Q_{V,x}$ the ideal, of $Q_{V,x}$, which is generated by the image of M_x under the canonical homomorphism $K[V] \to Q_{V,x}$. In terms of this notation our remarks prove

Theorem 19. _The non-units of $Q_{V,x}$ form the ideal $M_x Q_{V,x}$. This ideal is the kernel of the homomorphism (3.3.3) and therefore it is a K-rational maximal ideal. It is the only maximal ideal of $Q_{V,x}$ rational or otherwise._

We next examine briefly the connection between K-morphisms and localization. To this end we assume that $\phi : V \to W$ is a K-morphism of affine sets; and for $x \in V$, $y \in W$ we use M_x respectively N_y to denote the corresponding K-rational maximal ideal of $K[V]$ respectively $K[W]$.

The K-morphism ϕ induces a homomorphism $\phi^* : K[W] \to K[V]$ of K-algebras. Since, by (2.6.3),

$$\phi^*(N_{\phi(x)}) \subseteq M_x, \tag{3.3.6}$$

ϕ^* induces a homomorphism

$$\phi_x^* : Q_{W,\phi(x)} \to Q_{V,x} \tag{3.3.7}$$

of K-algebras. To see how the homomorphism operates, let $\theta' = f'/g'$, where $f', g' \in K[W]$ and $g' \notin N_{\phi(x)}$. Then

$$\phi_x^*(\theta') = \frac{\phi^*(f')}{\phi^*(g')} = \frac{f' \circ \phi}{g' \circ \phi}. \tag{3.3.8}$$

Accordingly

$$(\phi_x^*(\theta'))(x) = \theta'(\phi(x)) \tag{3.3.9}$$

and

$$\phi_x^*(N_{\phi(x)} Q_{W,\phi(x)}) \subseteq M_x Q_{V,x}. \tag{3.3.10}$$

In particular the inverse image, in (3. 3. 7), of the maximal ideal of $Q_{V,x}$ is the maximal ideal of $Q_{W,\phi(x)}$, a fact which is sometimes described by saying that ϕ_x^* is a <u>local homomorphism.</u>

Clearly if $\phi : V \to W$ is a K-isomorphism of affine sets, then ϕ_x^* is an isomorphism of K-algebras. Also if $\phi : V \to W$ and $\psi : W \to T$ are both of them K-morphisms of affine sets, then

$$(\psi \circ \phi)_x^* = \phi_x^* \circ \psi_{\phi(x)}^*. \qquad (3.3.11)$$

Finally let U be a closed subset of V and let $x \in U$. The inclusion K-morphism $j : U \to V$ induces a surjective homomorphism $j^* : K[V] \to K[U]$ of K-algebras whose kernel is $I_V(U)$. It follows that

$$j_x^* : Q_{V,x} \to Q_{U,x}$$

is surjective and its kernel is the ideal $I_V(U)Q_{V,x}$ that is generated by the image of $I_V(U)$ under the canonical homomorphism $K[V] \to Q_{V,x}$. Accordingly we obtain

Theorem 20. <u>Let</u> U <u>be a closed subset of the affine set</u> V <u>and let</u> $x \in U$. <u>Then the</u> K-<u>algebras</u> $Q_{U,x}$ <u>and</u> $Q_{V,x}/I_V(U)Q_{V,x}$ <u>are naturally isomorphic.</u>

3.4 Rational functions and dimension

Throughout section (3. 4), V will denote an affine set defined over K.

First suppose that V is irreducible. By Theorem 9, K[V] is an integral domain and therefore it has a quotient field. The quotient field will be denoted by K(V). It is, of course, an extension field of K.

Let $\theta \in K(V)$ and suppose that $x \in V$. We say that θ <u>is defined at</u> x if there exist f, $g \in K[V]$ such that $\theta = f/g$ and $g(x) \neq 0$. If N denotes the set where θ is defined, then N is a union of open sets and therefore it is itself an open set.

Suppose now that $x \in N$ so that there exist f, $g \in K[V]$ such that $\theta = f/g$ and $g(x) \neq 0$. We may put $\theta(x) = f(x)/g(x)$ because $f(x)/g(x)$ does not depend on the choice of f and g. In this way, we derive, from

θ, a function whose domain is N and which takes values in K. By an abuse of language a function obtained in this way is called a rational function on V.

Theorem 21. *Let V be an irreducible affine set and let θ_1, θ_2 belong to K(V). Then $\theta_1 = \theta_2$ if and only if there is a non-empty open subset N, of V, on which the corresponding rational functions are defined and equal.*

Proof. Assume that $\theta_1(y) = \theta_2(y)$ for all $y \in N$, where N is a non-empty open subset of V. Choose $x \in N$. Then there exist f_1, g_1, f_2, g_2 in K[V] such that $g_1(x) \neq 0$, $g_2(x) \neq 0$ and $f_i/g_i = \theta_i$. Next there is a non-empty open subset, N' say, of V on which $f_1g_2 - f_2g_1$ vanishes, and now it follows, by Theorem 14, that $f_1g_2 - f_2g_1 = 0$. Thus $\theta_1 = \theta_2$. This proves part of the theorem and the rest is trivial.

Theorem 21 shows that there is a natural bijection between K(V) and the set of rational functions on V. Indeed on the basis of this we may regard the rational functions, on the irreducible affine set V, as forming an extension field of K. In practice we often speak of the elements of K(V) as though they were rational functions thereby making the obvious identification.

Still assuming that V is irreducible, we note that the rational functions that are defined at the point x form a subalgebra of K(V), and that this subalgebra is none other than the local ring $Q_{V,x}$ of V at x.

We are now ready to introduce the notion of *dimension*.

Definition. The dimension, Dim V, of the irreducible affine set V is the transcendence degree of the field K(V), of rational functions on V, over the ground field K.

Now suppose that V is not necessarily irreducible and let $V_1, V_2, \ldots, V_p$ be its irreducible components. We extend the above definition by putting†

$$\text{Dim } V = \max\{\text{Dim } V_1, \text{Dim } V_2, \ldots, \text{Dim } V_p\}. \tag{3.4.1}$$

† The empty affine set is given the conventional dimension -1.

The next result is a crucial lemma in dimension theory. In order to prevent the statement of it from becoming too long, we preface it by some remarks which describe the general setting.

Suppose that R and S are finitely generated K-algebras, which are also integral domains. Denote by L the quotient field of R and by Ω that of S. Then L has transcendence degree, p say, over K and Ω has transcendence degree, q say, also over K. The result in question can now be stated as

Lemma 4. *Let the situation be as described above and let* $\phi: R \to S$ *be a surjective homomorphism of K-algebras. Then* $p \geq q$. *If* $p = q$, *then* ϕ *is an isomorphism.*

Proof. Let $R = K[\xi_1, \xi_2, \ldots, \xi_n]$ and put $\eta_i = \phi(\xi_i)$ so that $S = K[\eta_1, \eta_2, \ldots, \eta_n]$. Without loss of generality we may suppose that $\eta_1, \eta_2, \ldots, \eta_q$ are algebraically independent over K. Then $\xi_1, \xi_2, \ldots, \xi_q$ are also algebraically independent over K and therefore $p \geq q$. From here on we assume that $p = q$.

Suppose that $z \in R$ and $\phi(z) = 0$. It will suffice to show that z is zero. Let $Z, X_1, X_2, \ldots, X_q$ be indeterminates. Since z is algebraic over $K(\xi_1, \xi_2, \ldots, \xi_q)$, we can find a non-null member $F(Z, X_1, X_2, \ldots, X_q)$ of $K[Z, X_1, X_2, \ldots, X_q]$ such that $F(z, \xi_1, \xi_2, \ldots, \xi_q) = 0$ and, moreover, we can arrange that $F(Z, X_1, X_2, \ldots, X_q)$ is irreducible. Next $F(0, \eta_1, \eta_2, \ldots, \eta_q) = 0$ and therefore $F(0, X_1, X_2, \ldots, X_q) = 0$. Accordingly $F(Z, X_1, X_2, \ldots, X_q)$ has Z as a factor and thus we see that

$$F(Z, X_1, X_2, \ldots, X_q) = cZ$$

for some $c \neq 0$ belonging to K. This shows that $z = 0$ and completes the proof.

We are now ready to prove

Theorem 22. *Let* V *be an irreducible affine set and let* U *be a closed subset of* V. *If now* $U \neq V$, *then* $\mathrm{Dim}\, U < \mathrm{Dim}\, V$.

Proof. Clearly we may assume that U, as well as V, is irreducible. Consider the surjective homomorphism $K[V] \to K[U]$ that arises from the inclusion mapping of U into V. This has kernel $I_V(U)$ and, since $U \neq V$, $I_V(U) \neq 0$. The theorem therefore follows from Lemma 4.

Corollary. *Let* U *be a closed subset of the affine set* V. *Then* $\text{Dim } U \leq \text{Dim } V$.

Proof. The assertion is clear as soon as one notes that each component of U is contained in a component of V.

A second application of Lemma 4 is provided by

Theorem 23. *Let* V *be an irreducible affine set and* ξ *a generalized point of* V. *Then the transcendence degree of* $K(\xi)$ *over* K *is at most* Dim V. *Furthermore there is equality if and only if* ξ *is a generic point of* V.

Proof. If ξ is a generic point of V, then K(V) and $K(\xi)$ are isomorphic and therefore the transcendence degree of $K(\xi)$ over K is Dim V.

Now assume that ξ is not a generic point of V. By (2.11.1), there is a surjective homomorphism $K[V] \to K[\xi]$ of K-algebras and, by hypothesis, this is not an isomorphism. Thus, by Lemma 4, the transcendence degree of $K(\xi)$ over K is smaller than Dim V and so the theorem is proved.

We take this opportunity to identify those affine sets that have dimension zero.

Theorem 24. *The affine set* V *is irreducible and of dimension zero if and only if it consists of a single point. When this is the case* $K[V] = K$.

Proof. Assume that V is irreducible and that $\text{Dim } V = 0$. Since K[V] is a ring between K and K(V), and K(V) is an algebraic extension of K, K[V] itself must be a field and therefore it has exactly one maximal ideal namely (0). But K[V] has at least one K-rational maximal ideal.

This shows that $K[V] = K$ and that V has only one point. The remaining assertions of the theorem are trivial.

Corollary. Let V be a non-empty affine set. Then $\mathrm{Dim}\, V = 0$ if and only if V is a finite set.

There is another special case which it is convenient to mention at this point. Suppose that V is an n-dimensional vector space over K. If K is a finite field, then Theorem 24 Cor. shows that, as an affine set, V has dimension zero. On the other hand, if K is infinite, then $K[V]$ is isomorphic to the polynomial ring $K[X_1, X_2, \ldots, X_n]$ and therefore the dimension of V, as an affine set, is also n.

We next consider the dimension of the product of two affine sets. To this end let V, W be non-empty affine sets defined over K and, on the basis of (2.8.1), let us make the identification

$$K[V \times W] = K[V] \otimes_K K[W]. \tag{3.4.2}$$

The discussion given in section (1.5) shows that $K[V]$, $K[W]$ may be regarded as subalgebras of $K[V] \otimes_K K[W]$ and hence of $K[V \times W]$. On this understanding a typical element of $K[V \times W]$ may be written in the form

$$f_1 g_1 + f_2 g_2 + \ldots + f_m g_m, \tag{3.4.3}$$

where $f_i \in K[V]$ and $g_i \in K[W]$. It follows that if $K[V] = K[\xi_1, \xi_2, \ldots, \xi_p]$ and $K[W] = K[\eta_1, \eta_2, \ldots, \eta_q]$, then

$$K[V \times W] = K[\xi_1, \ldots, \xi_p, \eta_1, \ldots, \eta_q]. \tag{3.4.4}$$

Lemma 5. Suppose that V, W are non-empty affine sets. Let $f_1, f_2, \ldots, f_m$ belong to $K[V]$ and $g_1, g_2, \ldots, g_m$ to $K[W]$. If now $f_1, f_2, \ldots, f_m$ are linearly independent over K and $f_1 g_1 + f_2 g_2 + \ldots + f_m g_m = 0$, then $g_i = 0$ for $i = 1, 2, \ldots, m$.

Proof. In view of (3.4.2) this is simply a special case of Chapter 1 Theorem 1.

After these preliminaries let us assume that V and W are irreducible. By Theorems 9 and 16, each of K[V], K[W], and K[V × W] is an integral domain. Also, using our previous notation, $K(V) = K(\xi_1, \xi_2, \ldots, \xi_p)$, $K(W) = K(\eta_1, \eta_2, \ldots, \eta_q)$, and

$$K(V \times W) = K(\xi_1, \ldots, \xi_p, \eta_1, \ldots, \eta_q) \tag{3.4.5}$$

Let Dim V = r and Dim W = s. We may suppose, without loss of generality, that $\xi_1, \xi_2, \ldots, \xi_r$ are algebraically independent over K and $\eta_1, \eta_2, \ldots, \eta_s$ are also algebraically independent over K. This ensures that $K(V \times W)$ is an algebraic extension of $K(\xi_1, \ldots, \xi_r, \eta_1, \ldots, \eta_s)$ and so it follows that $\mathrm{Dim}(V \times W) \le r + s$. Now the set formed by the power products $\xi_1^{\nu_1} \xi_2^{\nu_2} \ldots \xi_r^{\nu_r}$ is linearly independent over K and thus we see, from Lemma 5, that it is also linearly independent over K(W). In other words, $\xi_1, \xi_2, \ldots, \xi_r$ are algebraically independent over $K(\eta_1, \eta_2, \ldots, \eta_q)$ and therefore $\mathrm{Dim}(V \times W) \ge r + s$. Our inequalities can now be combined to yield $\mathrm{Dim}(V \times W) = r + s$.

Theorem 25. Let V and W be non-empty affine sets defined over K. Then

$$\mathrm{Dim}(V \times W) = \mathrm{Dim}\, V + \mathrm{Dim}\, W.$$

Proof. In view of Theorem 17 we may assume that V and W are irreducible. However in that case the desired result follows from the preceding discussion.

3.5 Enlargement of the ground field

In section (3.5) we shall consider an affine set V, defined over K, along with an extension field L of K. As in section (2.10), this allows us to construct, in a natural manner, an affine set $V^{(L)}$, defined over L and containing V as a subset. Each member of K[V] has a natural prolongation to a member of $L[V^{(L)}]$ and, by identifying each member of K[V] with its prolongation, we may embed K[V] in $L[V^{(L)}]$. If this is done, then

$$L[V^{(L)}] = K[V]^L, \tag{3.5.1}$$

where the right-hand side is to be understood in the sense explained in section (1.6).

Theorem 26. <u>If V is irreducible, then so too is</u> $V^{(L)}$. <u>Furthermore V and $V^{(L)}$ have the same dimension.</u>

Proof. Suppose that $\omega, \omega' \in L[V^{(L)}]$ and $\omega\omega' = 0$. We shall show that either $\omega = 0$ or $\omega' = 0$. Now there exist $\lambda_1, \lambda_2, \ldots, \lambda_s$, in L and linearly independent over K, so that

$$\omega = \lambda_1 f_1 + \lambda_2 f_2 + \ldots + \lambda_s f_s,$$

$$\omega' = \lambda_1 f'_1 + \lambda_2 f'_2 + \ldots + \lambda_s f'_s,$$

where f_i and f'_i belong to $K[V]$. Denote by U respectively U' the closed subset of V where $f_1, f_2, \ldots, f_s$ respectively $f'_1, f'_2, \ldots, f'_s$ all vanish. Then when $x \in V$ either

$$\lambda_1 f_1(x) + \lambda_2 f_2(x) + \ldots + \lambda_s f_s(x) = 0$$

or

$$\lambda_1 f'_1(x) + \lambda_2 f'_2(x) + \ldots + \lambda_s f'_s(x) = 0.$$

This shows that $x \in U$ or $x \in U'$. Thus $V = U \cup U'$ whence $U = V$ or $U' = V$. If $U = V$, then all the f_i are zero and therefore $\omega = 0$. If $U' = V$, then $\omega' = 0$. Accordingly $V^{(L)}$ is irreducible.

Next let $K[V] = K[\xi_1, \xi_2, \ldots, \xi_p]$ and suppose that $\mathrm{Dim}\, V = r$. Without loss of generality we can assume that $\xi_1, \xi_2, \ldots, \xi_r$ are algebraically independent over K. By (3.5.1), $L[V^{(L)}] = L[\xi_1, \xi_2, \ldots, \xi_p]$ and this shows immediately that $\mathrm{Dim}\, V^{(L)} \leq r$. On the other hand, the power products $\xi_1^{\nu_1} \xi_2^{\nu_2} \ldots \xi_r^{\nu_r}$ are linearly independent over K and therefore by (3.5.1), linearly independent over L. Consequently we also have $\mathrm{Dim}\, V^{(L)} \geq r$ and now the proof is complete.

Theorem 27. <u>Let $V_1, V_2, \ldots, V_r$ be the irreducible components of the affine set</u> V. <u>Then</u> $V_1^{(L)}, V_2^{(L)}, \ldots, V_r^{(L)}$ <u>are the irreducible com</u>

ponents of $V^{(L)}$.

Remark. Here we use Chapter 2 Theorem 30. This allows us to regard $V_j^{(L)}$ as a closed subset of $V^{(L)}$.

Proof. By Theorem 26, $V_j^{(L)}$ is irreducible and, by Chapter 2 Theorem 30 Cor. 3,

$$V^{(L)} = V_1^{(L)} \cup V_2^{(L)} \cup \ldots \cup V_r^{(L)}.$$

Moreover Chapter 2 Theorem 30 Cor. 2 shows that none of $V_1^{(L)}, V_2^{(L)}, \ldots, V_r^{(L)}$ is contained in any of the others. The theorem follows.

Corollary 1. The relation $\operatorname{Dim} V = \operatorname{Dim} V^{(L)}$ holds even if the affine set V is reducible.

Corollary 2. Let ξ be a generalized point of V. Then the transcendence degree of $K(\xi)$ over K is at most $\operatorname{Dim} V$.

This corollary is obtained by combining Theorem 27 with Theorem 23.

At this point we shall interrupt the discussion of affine sets in order to introduce certain ideas that belong to the theory of fields.

Lemma 6. Let L_1, L_2 be subfields of a field Ω and let K be a common subfield of L_1 and L_2. Assume that every set of elements of L_1 which is linearly independent over K is also linearly independent over L_2. Then every set of elements of L_2 which is linearly independent over K is linearly independent over L_1.

Proof. Let R be the subring of Ω that is generated by L_1 and L_2. Then R is a K-algebra and from our assumption it follows that we have an isomorphism $L_1 \otimes_K L_2 \approx R$ which matches $\lambda_1 \otimes \lambda_2$ with $\lambda_1 \lambda_2$. (Here λ_i belongs to L_i.) The lemma therefore follows from Chapter 1 Theorem 1.

It will be noticed that, in the situation described in Lemma 6, the relation between L_1 and L_2 is a symmetrical one. This relation is described by saying that L_1 and L_2 are linearly disjoint over K.

Let Ω be an extension field of K and let $\overline{\Omega}$ be the algebraic closure of Ω. Then $\overline{\Omega}$ contains the algebraic closure $\overline{K}$ of K.

Definition. Suppose that, in the above situation, Ω and $\overline{K}$ are linearly disjoint over K. Then Ω is said to be a 'regular extension' of K.

Later we shall be concerned with the properties of regular extension For the moment, however, we merely note the definition and show that regular extensions arise naturally in connection with irreducible affine sets.

Suppose that V is irreducible. Then not only is K[V] an integral domain but, by Theorem 26, $L[V^{(L)}]$ is an integral domain as well. Now K[V] is a subring of $L[V^{(L)}]$. Consequently K(V) and L have $L(V^{(L)})$ as a common extension field. They also have K as a common subfield.

Theorem 28. Let V be irreducible. Then K(V) and L, considered as subfields of $L(V^{(L)})$, are linearly disjoint over K.

Proof. Let $f_1, f_2, \ldots, f_m$ belong to K(V) and be linearly independent over K. Further, suppose that $\lambda_1 f_1 + \lambda_2 f_2 + \ldots + \lambda_m f_m =$ where $\lambda_i \in L$. We have to show that $\lambda_1, \lambda_2, \ldots, \lambda_m$ are all zero. By multiplying $f_1, f_2, \ldots, f_m$ by a suitable non-zero element of K[V], we may suppose that $f_i \in K[V]$ for all i. But now what we have to prove is clear because, by (3.5.1), $L[V^{(L)}] = K[V]^L$.

Theorem 29. Let the affine set V be irreducible. Then K(V) is a regular extension of K.

This follows from Theorem 28 by taking L to be the algebraic closure of K.

We next examine the effect produced by an extension of the ground field on rational functions. As a preliminary to this examination we prove

Lemma 7. Let V be an irreducible affine set and O a non-empty open subset of V. Then the closure of O in $V^{(L)}$ is $V^{(L)}$ itself.

Proof. Theorem 2 shows that the closure of O in V is just V and we know, from Chapter 2 Theorem 29, that the closure of V in $V^{(L)}$ is $V^{(L)}$. The lemma follows from these facts.

Theorem 30. _Let the affine set_ V _be irreducible and let_ Θ_1, Θ_2 _be rational functions on_ $V^{(L)}$. _If_ Θ_1 _and_ Θ_2 _are defined and equal on some non-empty open subset of_ V, _then_ $\Theta_1 = \Theta_2$.

This is an easy consequence of Lemma 7.

Once again let V be irreducible. We know that K(V) is a subfield of $L(V^{(L)})$. Consequently each rational function on V gives rise to a rational function on $V^{(L)}$ or, as we shall say, each rational function on V has a _natural prolongation_ to a rational function on $V^{(L)}$. Let θ be a rational function on V and denote by Θ its prolongation to $V^{(L)}$. Considered as _functions_, θ is defined on an open subset of V and Θ on an open subset of $V^{(L)}$. Now assume that $x \in V$. At first sight it appears possible that Θ could be defined at x even though θ was not. However, as the following theorem shows, such inconvenient behaviour cannot, in fact, occur.

Theorem 31. _Suppose that the affine set_ V _is irreducible. Let_ θ _be a rational function on_ V _and denote by_ Θ _its natural prolongation to a rational function on_ $V^{(L)}$. _Finally let_ x _be a point of_ V. _Then_ Θ _is defined at_ x _if and only if_ θ _is defined at_ x.

Proof. It will be assumed that Θ is defined at x and we shall deduce that θ is defined at x. The converse is trivial.

By hypothesis there exist $f, g \in K[V]$ and $F, G \in L[V^{(L)}]$ such that $g \neq 0$, $G(x) \neq 0$, $F/G = \Theta$, and $\theta = f/g$. Thus $gF = fG$. Choose $\lambda_1, \lambda_2, \ldots, \lambda_n$ in L, linearly independent over K, so that

$$F = \lambda_1 f_1 + \lambda_2 f_2 + \ldots + \lambda_n f_n,$$

$$G = \lambda_1 g_1 + \lambda_2 g_2 + \ldots + \lambda_n g_n,$$

where the f_i and g_i belong to K[V]. Then

$$\sum_{i=1}^{n} \lambda_i(gf_i - fg_i) = 0$$

and therefore, because $L[V^{(L)}] = K[V]^L$, $gf_i = fg_i$ for all i. But $G(x) \neq 0$ and therefore we can choose i so that $g_i(x) \neq 0$. Since $\theta = f_i/g_i$ this proves that θ is defined at x.

3.6 Almost surjective K-morphisms

A number of results in section (3.6) require that the ground field K be algebraically closed. Whenever this is the case the fact is always made quite explicit in the statement of the result in question.

Definition. A K-morphism $\phi : V \to W$ of affine sets is said to be 'almost surjective' if the closure of $\phi(V)$ in W is W itself.

Theorem 32. The K-morphism $\phi : V \to W$ is almost surjective if and only if the associated homomorphism $\phi^* : K[W] \to K[V]$ of K-algebras is an injection.

Proof. If $\gamma \in K[W]$ and vanishes on $\phi(V)$, then it vanishes on the closure of $\phi(V)$. Consequently when ϕ is almost surjective ϕ^* must be an injection. We shall therefore assume that ϕ is not almost surjective and deduce that ϕ^* is not an injection.

In the situation under discussion the closure $\overline{\phi(V)}$, of $\phi(V)$, is a proper closed subset of W. Hence there exists $g \in K[W]$ such that the principal locus $C_W(g)$ contains $\phi(V)$ but is not the whole of W. But then

$$\phi^*(g) = g \circ \phi = 0$$

whereas $g \neq 0$. Thus the theorem is proved.

We now need some lemmas concerning commutative, finitely generated K-algebras, and for these it will be convenient to use a common notation. This will now be explained.

For Lemmas 8 to 10 inclusive R will denote a finitely generated K-algebra which is also an integral domain; and S will be a finitely generated subalgebra. By Dim R will be meant the transcendence degree of the quotient field of R over K, and if Π is a prime ideal of R we shall put

$$\dim \Pi = \operatorname{Dim}(R/\Pi).$$

Similar definitions apply to S and its prime ideals. Note that $\operatorname{Dim} S \le \operatorname{Dim} R$.

Since R is finitely generated as a K-algebra, we can certainly find $\xi_1, \xi_2, \ldots, \xi_n$ in R so that $R = S[\xi_1, \xi_2, \ldots, \xi_n]$. Put $S_0 = S$ and $S_i = S[\xi_1, \xi_2, \ldots, \xi_i]$ so that $S_n = R$ and $S_i = S_{i-1}[\xi_i]$. Of course, each S_i is a finitely generated subalgebra of R. We use $\mathfrak{M}(S_i)$ to denote the set of all maximal ideals of S_i. We recall that if K is <u>algebraically closed</u>, then $\mathfrak{M}(S_i)$ is the same as the set of all K-rational maximal ideals of S_i.

Lemma 8. *Suppose that* $f \in R$, $f \neq 0$. *Then there exists* $g \in S$, $g \neq 0$ *with the following property: given a prime ideal* P, *in* S, *such that* $g \notin P$ *there can be found a prime ideal* Π *in* R *for which* (i) $\Pi \cap S = P$, (ii) $f \notin \Pi$, *and* (iii) $\dim \Pi = \dim P + \operatorname{Dim} R - \operatorname{Dim} S$. *Such a* g *has the further property that if* $N \in \mathfrak{M}(S)$ *and* $g \notin N$, *then there exists* $M \in \mathfrak{M}(R)$ *satisfying* $f \notin M$ *and* $M \cap S = N$.

Proof. For the time being we shall ignore the final assertion. Suppose that for each i $(1 \le i \le n)$ we can establish the other assertions when R and S are replaced by S_i and S_{i-1} respectively. Evidently this part of the lemma will then follow. Accordingly we shall suppose that $R = S[\xi]$. The element f can now be expressed in the form

$$f = d_0 \xi^t + d_1 \xi^{t-1} + \ldots + d_t,$$

where $d_i \in S$ and $d_0 \neq 0$.

First assume that ξ is transcendental over the quotient field of S so that $\operatorname{Dim} R - \operatorname{Dim} S = 1$. In this instance we may take $g = d_0$. For if P is a prime ideal of S not containing d_0, then $\Pi = PS[\xi]$ is a prime ideal of R with the required properties.

Next assume that ξ is algebraic over the quotient field of S and therefore $\operatorname{Dim} R = \operatorname{Dim} S$. There exists $b \in S$, $b \neq 0$ such that $b\xi = \eta$

(say) is integral over S. Choose an integer $m > 0$ so that $b^m f = h$ (say) belongs to $S[\eta]$. Then we have a relation

$$h^q + c_1 h^{q-1} + c_2 h^{q-2} + \ldots + c_q = 0,$$

where $c_1, c_2, \ldots, c_q$ belong to S. Let us arrange that q is as small as possible. Then, because $h \neq 0$, we have $c_q \neq 0$.

Put $g = c_q b$. Then $g \in S$, $g \neq 0$. Now suppose that $g \notin P$, where P is a prime ideal of S. Because† $S[\eta]$ is an integral extension of S, there exists a prime ideal Q in $S[\eta]$ such that $Q \cap S = P$. Since $S[\eta]/Q$ is an integral extension of S/P, we have $\dim Q = \dim P$. Also $b \notin Q$ and $c_q \notin Q$. From $c_q \notin Q$ it follows that $h \notin Q$.

Denote by T the subring of the quotient field of $S[\eta]$ formed by all elements λ/b^ν, where $\lambda \in S[\eta]$ and $\nu \geq 0$ is an integer. Then

$$S[\eta] \subseteq R = S[\xi] \subseteq T$$

and QT is a prime ideal of T which contracts to Q in $S[\eta]$. Put $\Pi = QT \cap R$. This is a prime ideal of R which contracts to P in S. Also $f \notin \Pi$ because $h \notin \Pi$. Next

$$S[\eta]/Q \subseteq R/\Pi \subseteq T/QT$$

and the quotient field of $S[\eta]/Q$ is the same as the quotient field of T/QT. It follows that $\dim \Pi = \dim Q = \dim P$. Consequently Π has all the required properties, and (as already explained) the first part of the lemma follows in full generality.

Finally assume that $N \in \mathfrak{M}(S)$ and $g \notin N$. We know there exists a prime ideal Π, of R, such that $f \notin \Pi$ and $\Pi \cap S = N$. But Π is the intersection of all the members of $\mathfrak{M}(R)$ that contain it. Thus there exists $M \in \mathfrak{M}(R)$ for which $\Pi \subseteq M$ and $f \notin M$. But now $M \cap S = N$ because N is a _maximal_ ideal of S. This completes the proof.

Lemma 9. _Let_ Π _be a prime ideal of_ R _and_ P _a prime ideal of_ S. _Suppose that_ $\Pi \cap S = P$ _and that_ Π _is a minimal prime ideal of_

† See [(9) Theorem 8, p. 90].

PR. Then $\dim \Pi \geq \dim P + \operatorname{Dim} R - \operatorname{Dim} S$.

Proof. Put $q = \operatorname{Dim} S$ and $\nu = \dim P$. We can choose $x_1, x_2, \ldots, x_\nu$ in S so that their natural images in S/P are algebraically independent over K. Then $x_1, x_2, \ldots, x_\nu$ are themselves algebraically independent over K and, by forming fractions with respect to the non-zero elements of $K[x_1, x_2, \ldots, x_\nu]$ and replacing K by $K(x_1, x_2, \ldots, x_\nu)$, we can reduce the problem to the case where $\dim P=0$. Accordingly in what follows it will be assumed that P is a maximal ideal.

By the normalization theorem† there exist $q = \operatorname{Dim} S$ elements $\eta_1, \eta_2, \ldots, \eta_q$, in S, which are algebraically independent over K and are such that S is an integral extension of $K[\eta_1, \eta_2, \ldots, \eta_q]$. Then $P \cap K[\eta_1, \eta_2, \ldots, \eta_q]$ is a maximal ideal of $K[\eta_1, \eta_2, \ldots, \eta_q]$ and therefore it is possible to find q elements, say $\zeta_1, \zeta_2, \ldots, \zeta_q$, which generate it.‡ We claim that Π is a minimal prime ideal of $(\zeta_1, \zeta_2, \ldots, \zeta_q)R$. For let Π_0 be a prime ideal of R which satisfies

$$(\zeta_1, \zeta_2, \ldots, \zeta_q)R \subseteq \Pi_0 \subseteq \Pi.$$

Then $\Pi_0 \cap S$ and $\Pi \cap S$ are prime ideals of S which have the same contraction in $K[\eta_1, \eta_2, \ldots, \eta_q]$ and the former is contained in the latter. But this implies that

$$\Pi_0 \cap S = \Pi \cap S = P$$

and hence that $PR \subseteq \Pi_0 \subseteq \Pi$. Thus $\Pi = \Pi_0$ and our claim is established. But what we have just proved shows that§

$$\dim \Pi \geq \operatorname{Dim} R - q$$

and so the lemma follows.

---

† See [(13) Vol. 2 Theorem 25, p. 200].

‡ See [(9) Theorem 3, p. 281].

§ By [(9) Theorem 22, p. 217] we have rank $\Pi \leq q$, and rank $\Pi + \dim \Pi = \operatorname{Dim} R$ by [(13) Vol. 2 Theorem 20, p. 193].

Lemma 10. There exists $f \in R$, $f \neq 0$ with the following property: if Π is a prime ideal of R and $f \notin \Pi$, then $\dim \Pi \leq \dim(\Pi \cap S) + \operatorname{Dim} R - \operatorname{Dim} S$.

Proof. We continue to use the notation introduced just before Lemma 8 so that, in particular, $R = S[\xi_1, \xi_2, \ldots, \xi_n]$ and $S_i = S[\xi_1, \xi_2, \ldots, \xi_i]$. We use I to denote the set of integers i, between 1 and n, for which ξ_i is algebraic over the quotient field of S_{i-1}.

Let $i \in I$. There exists a non-null polynomial $p_i(X)$ with coefficie in S_{i-1} and such that $p_i(\xi_i) = 0$. We let f_i denote the leading coefficier of $p_i(X)$. Also if $1 \leq j \leq n$ and $j \notin I$ we put $f_j = 1$. Thus $f = f_1 f_2 \ldots f$ belongs to R and $f \neq 0$.

Now assume that Π is a prime ideal of R for which $f \notin \Pi$ and put $\Pi_t = \Pi \cap S_t$. Then $S_t/\Pi_t = (S_{t-1}/\Pi_{t-1})[\bar{\xi}_t]$, where $\bar{\xi}_t$ is the residue of ξ_t modulo Π_t, and therefore

$$\dim \Pi_t \leq \dim \Pi_{t-1} + 1.$$

Now suppose that $i \in I$. In this case the leading coefficient of $p_i(X)$ is not in Π_{i-1} and therefore $\bar{\xi}_i$ is algebraic over the quotient field of S_{i-1}/Π_{i-1}. Accordingly $\dim \Pi_i = \dim \Pi_{i-1}$. Thus to sum up

$$\dim \Pi_t - \dim \Pi_{t-1} = \begin{cases} 0 & \text{if } t \in I, \\ \leq 1 & \text{if } t \notin I. \end{cases}$$

The inequality $\dim \Pi - \dim(\Pi \cap S) \leq \operatorname{Dim} R - \operatorname{Dim} S$ follows by addition.

We shall now reformulate the substance of these lemmas in the language of affine sets. Note that if $\phi : V \to W$ is an almost surjective K-morphism of affine sets and V is irreducible, then W is irreducible as well. This follows from Theorems 1 and 4 and the fact that ϕ is continuous.

Theorem 33. Let $\phi : V \to W$ be an almost surjective K-morphism affine sets and suppose that V is irreducible and K is algebraically closed. If now U is a non-empty open subset of V, then $\phi(U)$ contains *a non-empty open subset of* W.

Proof. By Theorem 32, the associated homomorphism $\phi^* : K[W] \to K[V]$, of K-algebras, is an injection. Hence if we put $R = K[V]$, $S = K[W]$, then R is an integral domain and S is a subalgebra. Also $\text{Dim } R = \text{Dim } V$ and $\text{Dim } S = \text{Dim } W$. Next, because every closed subset of V is an intersection of principal loci, we can find $f \in R$, $f \neq 0$ such that

$$U_0 = \{x \mid x \in V \text{ and } f(x) \neq 0\}$$

is a non-empty open subset of V contained in U. Choose $g \in S$, $g \neq 0$ as in Lemma 8. Then

$$\{y \mid y \in W \text{ and } g(y) \neq 0\}$$

is a non-empty open subset of W and, by the final assertion of Lemma 8, it is contained in $\phi(U_0)$. Since $\phi(U_0) \subseteq \phi(U)$, the theorem is proved.

Corollary. *Suppose that* $f \in K[V]$, *where* V *is an affine set and* K *is algebraically closed. If* f *takes infinitely many different values on* V, *then* $K \backslash f(V)$ *if a finite set.*

Proof. By replacing V by a suitable irreducible component, we may suppose that V itself is irreducible. Let us regard K as being affine 1-space. Then $f : V \to K$ is a K-morphism and, because $f(V)$ is infinite, it is almost surjective. By Theorem 33, $f(V)$ contains a non-empty open subset of K, and now the corollary follows.

Theorem 34. *Suppose that* $\phi : V \to W$ *is almost surjective, that* V *is irreducible and that* K *is algebraically closed. Suppose also that* U *is a non-empty open subset of* V. *In these circumstances there exists a non-empty open subset* T, *of* W, *with the following property: given any irreducible closed subset* Y, *of* W, *which meets* T *there is a closed irreducible subset* X, *of* V, *such that* X *meets* U, $\phi(X) \subseteq Y$, *and* $\text{Dim } X = \text{Dim } Y + \text{Dim } V - \text{Dim } W$.

Proof. Put $R = K[V]$, $S = K[W]$ and regard S as a subalgebra of R. Choose $f \in K[V]$, $f \neq 0$ so that

$$U_0 = \{x \mid x \in V \text{ and } f(x) \neq 0\}$$

is contained in U. Without loss of generality we may suppose that $U = U$
in the subsequent discussion.

Choose $g \in S$, $g \neq 0$ as in Lemma 8 and put

$$T = \{y \mid y \in W \text{ and } g(y) \neq 0\}.$$

Then T is a non-empty open subset of W. Now assume that Y is an irreducible closed subset of W which meets T.

To Y corresponds a prime ideal P in S and we have $g \notin P$ because Y meets T. Consequently, by Lemma 8, there exists a prime ideal Π, of R, such that $f \notin \Pi$, $\Pi \cap S = P$, and

$$\dim \Pi = \dim P + \operatorname{Dim} V - \operatorname{Dim} W.$$

Since K is algebraically closed, there is a closed irreducible subset X, of V, which corresponds to Π. This has the required properties.

Theorem 35. *Let $\phi : V \to W$ be an almost surjective K-morphism where V is irreducible and K is algebraically closed. Let X be an irreducible closed subset of V and Y an irreducible closed subset of W If now X is a component of $\phi^{-1}(Y)$ and $\overline{\phi(X)} = Y$, then*

$$\operatorname{Dim} X \geq \operatorname{Dim} Y + \operatorname{Dim} V - \operatorname{Dim} W.$$

Proof. Once again put $R = K[V]$, $S = K[W]$ and regard S as a subalgebra of R. To X there will correspond a prime ideal Π in R and to Y a prime ideal P in S. Put $P' = \Pi \cap S$. Then, because $\phi(X) \subseteq Y$, we have $P \subseteq P'$. Let Y' be the irreducible closed subset of W that corresponds to P'. Then $\phi(X) \subseteq Y' \subseteq Y$, whence $Y' = Y$ becau $\overline{\phi(X)} = Y$. Thus $\Pi \cap S = P$ and hence $PR \subseteq \Pi$.

Suppose that Π' is a prime ideal of R satisfying $PR \subseteq \Pi' \subseteq \Pi$ and let X' be the irreducible closed subset of V that corresponds to Π Then $X \subseteq X'$ and $\phi(X') \subseteq Y$ because $\Pi' \cap S = P$. It follows that $X = X$

and therefore $\Pi' = \Pi$. Accordingly Π is a minimal prime ideal of PR and so

$$\dim \Pi \geq \dim P + \operatorname{Dim} R - \operatorname{Dim} S$$

by Lemma 9. But this is equivalent to

$$\operatorname{Dim} X \geq \operatorname{Dim} Y + \operatorname{Dim} V - \operatorname{Dim} W$$

and now the theorem is proved.

Theorem 36. Let $\phi : V \to W$ be an almost surjective K-morphism, where V is irreducible. Then there exists a non-empty open subset U, of V, such that if X is any irreducible closed subset of V which meets U, then $\operatorname{Dim} X \leq \operatorname{Dim}(\overline{\phi(X)}) + \operatorname{Dim} V - \operatorname{Dim} W$.

Proof. We put $R = K[V]$, $S = K[W]$ and regard S as a subalgebra of R. Choose $f \in R$, $f \neq 0$ as in Lemma 10 and set $U = \{x \mid x \in V$ and $f(x) \neq 0\}$.

Now suppose that X is an irreducible closed subset of V that meets U. To X there will correspond a prime ideal Π of R and, since X meets U, we have $f \notin \Pi$. Next, to $\overline{\phi(X)}$ corresponds a prime ideal P, of S, and each rational maximal ideal of R that contains Π contracts to a rational maximal ideal of S that contains P. But (Theorem 12) Π is the intersection of all the rational maximal ideals that contain it. Consequently $P \subseteq \Pi \cap S$ and therefore, by Lemma 10,

$$\begin{aligned} \dim \Pi &\leq \dim(\Pi \cap S) + \operatorname{Dim} R - \operatorname{Dim} S \\ &\leq \dim P + \operatorname{Dim} V - \operatorname{Dim} W \end{aligned}$$

and so the theorem is proved.

Theorem 37. Let $\phi : V \to W$ be an almost surjective K-morphism of affine sets, where V is irreducible and K is algebraically closed. In these circumstances there exists a non-empty open subset T, of W, with the following property: for any closed irreducible subset Y (of W) which meets T, $\phi^{-1}(Y)$ has a component whose dimension is equal to

$\text{Dim } Y + \text{Dim } V - \text{Dim } W$.

Proof. Choose a non-empty open subset U, of V, as in Theorem 36 and then a non-empty open subset T, of W, as in Theorem 34. Now suppose that Y is a closed irreducible subset of W which meets T.

By Theorem 34, there exists a closed irreducible subset X, of V, which meets U, is contained in $\phi^{-1}(Y)$, and satisfies

$$\text{Dim } X = \text{Dim } Y + \text{Dim } V - \text{Dim } W.$$

Choose a component X' of $\phi^{-1}(Y)$ so that $X \subseteq X'$. Then X' meets U and therefore, by Theorem 36,

$$\begin{aligned} \text{Dim } X' &\leq \text{Dim}(\overline{\phi(X')}) + \text{Dim } V - \text{Dim } W \\ &\leq \text{Dim } Y + \text{Dim } V - \text{Dim } W \\ &= \text{Dim } X. \end{aligned}$$

Accordingly, by Theorem 22, $X = X'$ and the theorem is proved.

Corollary. Suppose that $y \in T$. Then $\phi^{-1}(\{y\})$ has a component whose dimension is equal to $\text{Dim } V - \text{Dim } W$.

It will be recalled, from section (2. 6), that a K-morphism can be a bijection without being a K-isomorphism. Theorem 33 will now be used to throw some light on this situation.

Lemma 11. Let $\phi : V \to W$ be an almost surjective K-morphism which is also an injection. Assume that V and W are irreducible and that K is algebraically closed. Then $\text{Dim } V = \text{Dim } W$.

Proof. By hypothesis $K[V]$ is an integral domain and $K[W]$ may be regarded as a subalgebra. We wish to show that $K(V)$ is an algebrai extension of $K(W)$. We shall assume that it is not and seek a contradict

Our assumptions ensure that there exists $T \in K(V)$ which is trans- cendental over $K(W)$; indeed we can choose T so that it lies in $K[V]$. Then

$$K[W] \subseteq (K[W])[T] \subseteq K[V].$$

Next $(K[W])[T]$ may be regarded as the coordinate ring of $W \times K$ and then the inclusion homomorphisms $K[W] \to (K[W])[T]$ and $(K[W])[T] \to K[V]$ correspond respectively to the projection $\pi : W \times K \to W$ and a certain K-morphism $\psi : V \to W \times K$. Furthermore $\phi = \pi \circ \psi$.

By Theorem 32, ψ is almost surjective and therefore, by Theorem 33, $\psi(V)$ contains a non-empty open subset U of $W \times K$. Let (w_0, k_0) belong to U and consider the K-morphism $K \to W \times K$ in which $k \mapsto (w_0, k)$. The inverse image of U is a non-empty open subset of K and therefore it contains infinitely many elements. In particular we can find $k_1 \in K$ so that $k_1 \neq k_0$ and $(w_0, k_1) \in U$. Thus (w_0, k_0) and (w_0, k_1) are distinct points of $\psi(V)$ having the same projection on to W. This, however, contradicts our original assumption that $\phi = \pi \circ \psi$ is an injection.

The hypotheses of the lemma can be relaxed a little. This is shown by

Theorem 38. _Let_ $\phi : V \to W$ _be an almost surjective_ K-_morphism of affine sets which is also an injection. If now_ K _is algebraically closed, then_ $\operatorname{Dim} V = \operatorname{Dim} W$.

Proof. Let $V_1, V_2, \ldots, V_r$ be the irreducible components of V (we may assume that V is non-empty) and let $\overline{\phi(V_j)}$ denote the closure of $\phi(V_j)$ in W. Then ϕ induces an almost surjective K-morphism

$$\phi_j : V_j \to \overline{\phi_j(V_j)}$$

which is also an injection. Consequently, by Lemma 11,

$$\operatorname{Dim} V_j = \operatorname{Dim}(\overline{\phi(V_j)}).$$

Now $\overline{\phi(V_1)}, \overline{\phi(V_2)}, \ldots, \overline{\phi(V_r)}$ are closed irreducible subsets of W and, because $\overline{\phi(V)} = W$, their union is W. Accordingly the irreducible components of W are the maximal members of

$$\{\overline{\phi(V_1)}, \overline{\phi(V_2)}, \ldots, \overline{\phi(V_r)}\}$$

and therefore it follows, in view of Theorem 22, that

$$\begin{aligned}\operatorname{Dim} W &= \max_{1\le j\le r} \operatorname{Dim}(\overline{\phi(V_j)}) \\ &= \max_{1\le j\le r} \operatorname{Dim} V_j \\ &= \operatorname{Dim} V.\end{aligned}$$

4 · Derivations and tangent spaces

General remarks

In this chapter the concept of a derivation is introduced. This will provide an important new tool which is useful in studying local properties of affine sets.

As before K denotes a field. Unless there is an explicit statement to the contrary, no special assumptions are made concerning K.

Suppose that V is an affine set, defined over K, and that x is a point of V. As in section (3. 3) we use $Q_{V,x}$ to denote the local ring of V at x. By Chapter 3 Theorem 19, $Q_{V,x}$ has only one maximal ideal. In this chapter the maximal ideal is denoted by $M_{V,x}$.

4. 1 Derivations in algebras

Throughout section (4. 1) we use R, S, T to denote K-algebras. It will be assumed that each of them is unitary, associative and commutative.

Let $\phi : R \to S$ be a homomorphism of K-algebras. By a K-derivation of type ϕ, of R into S, we shall understand a K-linear mapping $D : R \to S$ with the property that

$$D(xy) = (Dx)\phi(y) + \phi(x)(Dy) \tag{4. 1. 1}$$

for all x, y in R. The set of all such derivations of R into S will be denoted by $\mathrm{Der}_K(R, S, \phi)$.

There are various situations where the notation can conveniently be simplified. For example, if R is a subalgebra of S we put

$$\mathrm{Der}_K(R, S) = \mathrm{Der}_K(R, S, j), \tag{4. 1. 2}$$

where $j : R \to S$ is the inclusion homomorphism. We also set

$$\mathrm{Der}_K(R) = \mathrm{Der}_K(R, R, j), \tag{4.1.3}$$

where, this time, $j : R \to R$ is the identity homomorphism.

Another case where the notation can be simplified arises as follows Let Q be a K-algebra (unitary, associative, and commutative) which has exactly one K-rational maximal ideal. In this case there is only a single possible homomorphism $Q \to K$ of K-algebras and therefore we can writ $\mathrm{Der}_K(Q, K)$ without risk of ambiguity.

Let us return to the general situation. If D_1, D_2 belong to $\mathrm{Der}_K(R, S, \phi)$ and $k \in K$, and if we define $D_1 + D_2$ and kD_1 in the obvious manner, then these too belong to $\mathrm{Der}_K(R, S, \phi)$. In fact we hav

Lemma 1. <u>Let</u> $\mathrm{Hom}_K(R, S)$ <u>be regarded as a vector space over</u> <u>Then</u> $\mathrm{Der}_K(R, S, \phi)$ <u>is a subspace.</u>

Assume that $D \in \mathrm{Der}_K(R, S, \phi)$ and $s \in S$. Define

$$sD : R \to S \tag{4.1.4}$$

by

$$(sD)x = s(Dx). \tag{4.1.5}$$

Then $sD \in \mathrm{Der}_K(R, S, \phi)$. Bearing this in mind we at once obtain

Lemma 2. <u>Let the notation be as above. Then (4.1.4) endows</u> $\mathrm{Der}_K(R, S, \phi)$ <u>with a natural structure as an</u> S-<u>module.</u>

It should be noted that the S-module structure of $\mathrm{Der}_K(R, S, \phi)$ is compatible with its structure as a K-space.

Lemma 3. <u>If</u> $D \in \mathrm{Der}_K(R, S, \phi)$, <u>then</u> $D(k1_R) = 0$ <u>for all</u> $k \in K$.

Proof. It is enough to show that $D1_R = 0$. This, however, follow from

$$\begin{aligned} D1_R &= D(1_R 1_R) \\ &= \phi(1_R)(D1_R) + (D1_R)\phi(1_R) \\ &= D1_R + D1_R. \end{aligned}$$

When R is non-trivial K may be regarded as a subfield of R. Lemma 3 then states that every K-derivation of R into S vanishes on K. In this connection we note

Lemma 4. *Suppose that R is non-trivial and that $D : R \to S$ is a mapping which satisfies*

$$D(x + y) = Dx + Dy,$$
$$D(xy) = (Dx)\phi(y) + \phi(x)(Dy),$$

for all $x, y \in R$. Then $D \in \mathrm{Der}_K(R, S, \phi)$ if and only if D vanishes on K.

Proof. Assume that $D(k) = 0$ for all $k \in K$. Then, for $x \in R$,

$$D(kx) = \phi(k)(Dx) + (Dk)\phi(x) = k(Dx)$$

and therefore $D \in \mathrm{Der}_K(R, S, \phi)$. The converse follows from Lemma 3.

It is sometimes necessary to decide when two derivations coincide. For this the following result is often useful.

Theorem 1. *Let $D, D' \in \mathrm{Der}_K(R, S, \phi)$ and let A be a subset of R which generates R as a K-algebra. If now $Da = D'a$ for all $a \in A$, then $D = D'$.*

Proof. Let $a_1, a_2, \ldots, a_n$ belong to A and let $k \in K$. It is clear that

$$D(ka_1a_2 \ldots a_n) = D'(ka_1a_2 \ldots a_n).$$

Since a typical element of R is a sum of elements such as $ka_1a_2 \ldots a_n$, the theorem follows.

The next lemma embodies the same idea as Theorem 1 but it is more explicit.

Lemma 5. *Let $R = K[\xi_1, \xi_2, \ldots, \xi_n]$ and suppose that $z = f(\xi_1, \xi_2, \ldots, \xi_n)$, where $f(X_1, X_2, \ldots, X_n)$ belongs to the polynomial ring $K[X_1, X_2, \ldots, X_n]$. Then*

$$Dz = \sum_{i=1}^{n} \phi\left(\frac{\partial f}{\partial \xi_i}\right) \quad (D\xi_i)$$

for any D *in* $\mathrm{Der}_K(R, S, \phi)$.

This is obvious.

It will now be shown how, from given derivations, we can sometime derive new ones. To this end let $\phi : R \to S$ and $\psi : S \to T$ be homomorphisms of K-algebras and let $D \in \mathrm{Der}_K(R, S, \phi)$. It is easily verified that $\psi \circ D$ belongs to $\mathrm{Der}_K(R, T, \psi \circ \phi)$.

Theorem 2. *Let the situation be as described above. Then the mapping*

$$\mathrm{Der}_K(R, S, \phi) \to \mathrm{Der}_K(R, T, \psi \circ \phi),$$

in which $D \mapsto \psi \circ D$, *is* K-*linear.*

We leave the verification to the reader.

Let $\phi : R \to S$ and $\psi : S \to T$ be as before and suppose that $\Delta \in \mathrm{Der}_K(S, T, \psi)$. It is easily checked that (i) $\Delta \circ \phi$ belongs to $\mathrm{Der}_K(R, T, \psi \circ \phi)$, (ii) the new derivation vanishes on $\mathrm{Ker}\ \phi$, and (iii) the mapping

$$\mathrm{Der}_K(S, T, \psi) \to \mathrm{Der}_K(R, T, \psi \circ \phi), \tag{4.1.6}$$

in which $\Delta \mapsto \Delta \circ \phi$, is K-linear.

Theorem 3. *Let the situation be as described in the preceding paragraph and suppose that the homomorphism* $\phi : R \to S$ *is surjective. Then the* K-*linear mapping*

$$\mathrm{Der}_K(S, T, \psi) \to \mathrm{Der}_K(R, T, \psi \circ \phi)$$

described in (4.1.6) *establishes a bijection between* $\mathrm{Der}_K(S, T, \psi)$ *and the subspace of* $\mathrm{Der}_K(R, T, \psi \circ \phi)$ *formed by the derivations which vani on* $\mathrm{Ker}\ \phi$.

Proof. The fact that ϕ is surjective obviously ensures that (4.1.6) is an injection. Now suppose that D belongs to $\mathrm{Der}_K(R, T, \psi \circ \phi)$ and that it vanishes on $\mathrm{Ker}\ \phi$. Then D gives rise to a K-linear mapping $\Delta : S \to T$ which is such that $\Delta(\phi(r)) = Dr$ for all $r \in R$. A simple verification shows that Δ is in $\mathrm{Der}_K(S, T, \psi)$ and, by construction, $D = \Delta \circ \phi$. Accordingly (4.1.6) is also a surjection and the proof is complete.

Our next result has to do with the formation of fractions. For the moment we assume that our K-algebras R, S are <u>integral domains</u>; in addition we suppose that Σ is a non-empty multiplicatively closed subset of R not containing zero, and that Ω is a non-empty multiplicatively closed subset of S also not containing the zero element. Denote by R_Σ the set of all elements belonging to the quotient field of R that can be expressed in the form r/σ, where $r \in R$ and $\sigma \in \Sigma$; and define S_Ω similarly. Evidently R_Σ and S_Ω may be regarded as K-algebras. Moreover R is a subalgebra of R_Σ and S a subalgebra of S_Ω.

Now suppose that $\phi : R \to S$ is a homomorphism of K-algebras and that it satisfies

$$\phi(\Sigma) \subseteq \Omega. \tag{4.1.7}$$

Then ϕ extends to a homomorphism

$$\bar{\phi} : R_\Sigma \to S_\Omega \tag{4.1.8}$$

of K-algebras in which, with a self-explanatory notation,

$$\bar{\phi}\left(\frac{r}{\sigma}\right) = \frac{\phi(r)}{\phi(\sigma)}. \tag{4.1.9}$$

Next if $D \in \mathrm{Der}_K(R, S, \phi)$, then it is possible to define a mapping

$$\bar{D} : R_\Sigma \to S_\Omega \tag{4.1.10}$$

by means of the formula

$$\bar{D}\left(\frac{r}{\sigma}\right) = \frac{\phi(\sigma)(Dr) - \phi(r)(D\sigma)}{[\phi(\sigma)]^2} \tag{4.1.11}$$

and a straightforward verification shows not only that $\overline{D} \in \mathrm{Der}_K(R_\Sigma, S_\Omega, \overline{\phi})$ but also that $\overline{D}$ agrees with D on R and is the only member of $\mathrm{Der}_K(R_\Sigma, S_\Omega, \overline{\phi})$ to do so. For future reference we record

Lemma 6. *Let* R, S, ϕ, Σ *and* Ω *be as above. Then the mapping*

$$\mathrm{Der}_K(R, S, \phi) \to \mathrm{Der}_K(R_\Sigma, S_\Omega, \overline{\phi}),$$

which results from extending D *in* $\mathrm{Der}_K(R, S, \phi)$ *by means of the formula (4.1.11), is K-linear and an injection.*

Corollary. *In the special case where* S *is a field the mapping*

$$\mathrm{Der}_K(R, S, \phi) \to \mathrm{Der}_K(R_\Sigma, S_\Omega, \overline{\phi}),$$

described in the lemma, is an isomorphism of K-spaces.

It should be noted that in this instance S_Ω coincides with S.

Once again let $\phi : R \to S$ be a homomorphism of K-algebras, but now let L be an extension field of K. Then, as in section (1.6), $\phi^L : R^L \to S^L$ is a homomorphism of L-algebras. Also if $D : R \to S$ is a K-linear mapping, then $D^L : R^L \to S^L$ is L-linear. Note that if $a \in R$ and $\lambda \in L$, then $\phi^L(a\lambda) = \lambda\phi(a)$ and $D^L(\lambda a) = \lambda(Da)$.

Lemma 7. *Suppose that* D *belongs to* $\mathrm{Der}_K(R, S, \phi)$. *Then* D^L *belongs to* $\mathrm{Der}_L(R^L, S^L, \phi^L)$.

Proof. Let $x, y \in R^L$. Then we can write

$$x = \lambda_1 a_1 + \lambda_2 a_2 + \ldots + \lambda_m a_m,$$
$$y = \lambda'_1 b_1 + \lambda'_2 b_2 + \ldots + \lambda'_n b_n,$$

where $\lambda_i, \lambda'_j \in L$ and $a_i, b_j \in R$. Next

$$\begin{aligned} D^L(xy) &= \sum_i\sum_j \lambda_i\lambda'_j D(a_i b_j) \\ &= \sum_i\sum_j \lambda_i(Da_i)\lambda'_j\phi(b_j) + \sum_i\sum_j \lambda_i\phi(a_i)\lambda'_j(Db_j) \\ &= (D^L x)\phi^L(y) + \phi^L(x)(D^L y). \end{aligned}$$

Since D^L is L-linear, this completes the proof.

4.2 Examples of derivations

We pause, for a moment, to illustrate some of the general ideas introduced in the last section. As before R, S denote K-algebras which are assumed to be unitary, associative, and commutative.

Example 1. Let $X_1, X_2, \ldots, X_n$ be indeterminates. Then, for each i $(1 \le i \le n)$,

$$\frac{\partial}{\partial X_i} : K[X_1, X_2, \ldots, X_n] \to K[X_1, X_2, \ldots, X_n]$$

belongs to $\mathrm{Der}_K(K[X_1, X_2, \ldots, X_n])$.

Example 2. Let $X_1, X_2, \ldots, X_n$ be indeterminates and suppose that

$$\psi : K[X_1, X_2, \ldots, X_n] \to S$$

is a homomorphism of K-algebras. By Example 1 and Theorem 2,

$$\Delta_i = \psi \circ \frac{\partial}{\partial X_i} \qquad (4.2.1)$$

is a member of $\mathrm{Der}_K(K[X_1, X_2, \ldots, X_n], S, \psi)$. Next let $\sigma_1, \sigma_2, \ldots, \sigma_n$ belong to S. Then

$$\Delta = \sigma_1\Delta_1 + \sigma_2\Delta_2 + \ldots + \sigma_n\Delta_n$$

belongs to $\mathrm{Der}_K(K[X_1, \ldots, X_n], S, \psi)$ and it is characterized by the property that

$$(\sigma_1\Delta_1 + \sigma_2\Delta_2 + \ldots + \sigma_n\Delta_n)X_i = \sigma_i \qquad (4.2.2)$$

for $i = 1, 2, \ldots, n$. This establishes

Theorem 4. The S-module $\mathrm{Der}_K(K[X_1, \ldots, X_n], S, \psi)$ is free and if Δ_i is defined as in (4.2.1), then $\Delta_1, \Delta_2, \ldots, \Delta_n$ form a base of this module.

As a matter of notation we recall that if $\psi(X_i) = \eta_i$, then it is usual to write

$$\psi \circ \frac{\partial}{\partial X_i} = \frac{\partial}{\partial \eta_i} . \tag{4.2.3}$$

Thus $\partial/\partial\eta_1, \partial/\partial\eta_2, \ldots, \partial/\partial\eta_n$ is a base for the S-module $\mathrm{Der}_K(K[X_1, \ldots, X_n], S, \psi)$.

Example 3. We next consider the case of a finitely generated K-algebra. Suppose then that

$$R = K[\xi_1, \xi_2, \ldots, \xi_n]. \tag{4.2.4}$$

Let $X_1, X_2, \ldots, X_n$ be indeterminates and denote by θ the surjective homomorphism

$$\theta : K[X_1, X_2, \ldots, X_n] \to K[\xi_1, \xi_2, \ldots, \xi_n] \tag{4.2.5}$$

of K-algebras in which $\theta(X_i) = \xi_i$. Put

$$\mathfrak{A} = \mathrm{Ker}\ \theta . \tag{4.2.6}$$

Then, since $K[X_1, X_2, \ldots, X_n]$ is a Noetherian ring, we can find a finite set of polynomials, say

$$F_j(X_1, X_2, \ldots, X_n) \text{ where } 1 \le j \le m,$$

which generates $\mathfrak{A}$.

Now suppose that $\phi : R \to S$ is a homomorphism of K-algebras and put

$$\phi(\xi_i) = x_i. \tag{4.2.7}$$

Then Example 2 shows that

$$\mathrm{Der}_K(K[X_1, \ldots, X_n], S, \phi \circ \theta)$$

is a free S-module having $\partial/\partial x_1, \partial/\partial x_2, \ldots, \partial/\partial x_n$ as a base. Hence if Δ belongs to

$$\mathrm{Der}_K(K[X_1, X_2, \ldots, X_n], S, \phi \circ \theta),$$

then there exist unique elements $\sigma_1, \sigma_2, \ldots, \sigma_n$ in S such that

$$\Delta = \sigma_1 \frac{\partial}{\partial x_1} + \sigma_2 \frac{\partial}{\partial x_2} + \ldots + \sigma_n \frac{\partial}{\partial x_n} \tag{4.2.8}$$

and, for every F in $K[X_1, X_2, \ldots, X_n]$,

$$\Delta F = \sigma_1 \frac{\partial F}{\partial x_1} + \sigma_2 \frac{\partial F}{\partial x_2} + \ldots + \sigma_n \frac{\partial F}{\partial x_n}.$$

Consider the equations

$$\left.\begin{aligned} \frac{\partial F_1}{\partial x_1}\sigma_1 + \frac{\partial F_1}{\partial x_2}\sigma_2 + \ldots + \frac{\partial F_1}{\partial x_n}\sigma_n &= 0 \\ \frac{\partial F_2}{\partial x_1}\sigma_1 + \frac{\partial F_2}{\partial x_2}\sigma_2 + \ldots + \frac{\partial F_2}{\partial x_n}\sigma_n &= 0 \\ \cdot \quad\quad \cdot \quad\quad \cdots & \\ \cdot \quad\quad \cdot \quad\quad \cdots & \\ \frac{\partial F_m}{\partial x_1}\sigma_1 + \frac{\partial F_m}{\partial x_2}\sigma_2 + \ldots + \frac{\partial F_m}{\partial x_n}\sigma_n &= 0 \end{aligned}\right\}. \tag{4.2.9}$$

Lemma 8. Suppose that

$$\Delta = \sigma_1 \frac{\partial}{\partial x_1} + \sigma_2 \frac{\partial}{\partial x_2} + \ldots + \sigma_n \frac{\partial}{\partial x_n},$$

where $\sigma_1, \sigma_2, \ldots, \sigma_n$ belong to S. Then Δ vanishes on $\mathfrak{A}$ if and only if the equations (4.2.9) are satisfied.

Proof. Each F in $\mathfrak{A}$ can be expressed in the form

$$F = G_1F_1 + G_2F_2 + \ldots + G_mF_m,$$

where $G_j \in K[X_1, X_2, \ldots, X_n]$. Since $F_j(x_1, x_2, \ldots, x_n) = 0$, we see that $\Delta F = 0$ for all $F \in \mathfrak{A}$ if and only if $\Delta F_j = 0$ for $j = 1, 2, \ldots, m$. The lemma follows.

In the next theorem $\sigma_1, \sigma_2, \ldots, \sigma_n$ continue to denote elements of S.

Theorem 5. _There is a bijection between the set of solutions of the equations (4.2.9) and_ $Der_K(R, S, \phi)$. _This is such that if_ $D \in Der_K(R, S, \phi)$, _then the corresponding solution of the equations (4.2.9) is given by_ $D\xi_i = \sigma_i$ $(1 \le i \le n)$.

In view of our previous remarks, this is an immediate consequence of Theorem 3.

These observations can be taken a stage further. We know that $Der_K(R, S, \phi)$ is an S-module. Now the solutions, in S, of the equations (4.2.9) also form an S-module. Let us call this the _solution module_ of the equations. We then have

Theorem 6. _The bijection of Theorem 5 is an isomorphism between the S-module_ $Der_K(R, S, \phi)$ _and the solution module of the equations (4.2.9) when these are regarded as equations to be solved in_ S.

Example 4. Let $X_1, X_2, \ldots, X_n$ continue to denote indeterminates and let L be an extension field of the quotient field $K(X_1, X_2, \ldots, X_n)$ of $K[X_1, X_2, \ldots, X_n]$. Then $K[X_1, X_2, \ldots, X_n]$ is a subalgebra of L and, by Theorem 4,

$$Der_K(K[X_1, X_2, \ldots, X_n], L)$$

is a vector space over L and as such it has the n derivations $\partial/\partial X_i$ as a base.

Next our discussion of derivations in relation to fractions shows that each member of $Der_K(K[X_1, \ldots, X_n], L)$ has a unique extension to a member of $Der_K(K(X_1, \ldots, X_n), L)$. The extension of $\partial/\partial X_i$ is denoted by the same symbol. We now see, from Lemma 6 Cor., that these n extended derivations form a base for $Der_K(K(X_1, \ldots, X_n), L)$ considered as a vector space over L.

4.3 Derivations in fields

We must now make a special study of derivations in the situation where the algebras involved are fields. Accordingly throughout section (4.3) L will denote an extension field of K.

Suppose that K^* is a field between K and L and that $D \in \mathrm{Der}_K(K^*, L)$. If now X is an indeterminate and $g(X) \in K^*[X]$, then by $g^D(X)$ will be understood the polynomial (with coefficients in L) that is obtained by applying D to the coefficients of $g(X)$. Note that if $g(X) = p(X)q(X)$, where $p(X)$, $q(X)$ are in $K^*[X]$, then

$$g^D(X) = p^D(X)q(X) + p(X)q^D(X). \tag{4.3.1}$$

Again the notation $g'(X)$ will be used for the formal derivative of $g(X)$. Hence, in addition to (4.3.1), we have

$$g'(X) = p'(X)q(X) + p(X)q'(X). \tag{4.3.2}$$

Let $\xi \in L$ and consider the possibility of extending D, in $\mathrm{Der}_K(K^*, L)$, so as to yield a member of $\mathrm{Der}_K(K^*(\xi), L)$.

Lemma 9. *Let K, K^*, L and D be as described above. Suppose that ξ belongs to L, is algebraic over K^*, and has $f(X)$ as its irreducible polynomial over K^*. Finally let $\lambda \in L$. Then in order that there should exist $\Delta \in \mathrm{Der}_K(K^*(\xi), L)$ which (i) extends D, and (ii) satisfies $\Delta\xi = \lambda$, it is necessary and sufficient that*

$$f^D(\xi) + f'(\xi)\lambda = 0. \tag{4.3.3}$$

If Δ exists, then it is unique.

Proof. If Δ exists, then, by applying it to the equation $f(\xi) = 0$, we immediately obtain

$$f^D(\xi) + f'(\xi)\lambda = 0.$$

From now on we shall assume that (4.3.3) holds.

Suppose that $h(X) \in K^*[X]$ and $h(\xi) = 0$. Then $h(X) = f(X)p(X)$ for some $p(X) \in K^*[X]$ and $h^D(\xi) = f^D(\xi)p(\xi)$, $h'(\xi) = f'(\xi)p(\xi)$. Accordingly

$$h^D(\xi) + h'(\xi)\lambda = 0.$$

It follows that if $g_1(X)$, $g_2(X)$ belong to $K^*[X]$ and $g_1(\xi) = g_2(\xi)$, then

$$g_1^D(\xi) + g_1'(\xi)\lambda = g_2^D(\xi) + g_2'(\xi)\lambda$$

and therefore we can define a mapping $\Delta : K^*[\xi] \to L$ by means of the formula

$$\Delta(g(\xi)) = g^D(\xi) + g'(\xi)\lambda.$$

But $K^*[\xi] = K^*(\xi)$. Also it is a simple matter to verify that $\Delta \in \mathrm{Der}_K(K^*(\xi), L)$ and satisfies conditions (i) and (ii). Finally it is obvious that at most one member of $\mathrm{Der}_K(K^*(\xi), L)$ can have these properties.

Corollary 1. Let K, K^*, L, D be as above and suppose that $\xi \in L$ and is algebraic over K^*. Then the following statements are equivalent:

(a) D has exactly one extension which is a member of $\mathrm{Der}_K(K^*(\xi), L)$;

(b) ξ is separable over K^*.

Proof. We use the same notation as in the proof of the lemma.

The number of extensions of D is the same as the number of solutions of

$$f^D(\xi) + f'(\xi)\lambda = 0,$$

where λ has to belong to L. Hence if ξ is separable over K^*, that is if $f'(\xi) \neq 0$, then D has exactly one extension. However, if $f'(\xi) = 0$, then there may be no solution; but if there is a solution, then there will be more than one.

Corollary 2. Let $\xi \in L$ and be algebraic over K. Then the L-space $\mathrm{Der}_K(K(\xi), L)$ has dimension zero if ξ is separable over K, whereas it has dimension one if ξ is not separable over K.

Proof. In Corollary 1 take $K^* = K$. Certainly $\mathrm{Der}_K(K, L) = 0$ and if ξ is separable over K, then the zero derivation of K into L has only the zero extension in $\mathrm{Der}_K(K(\xi), L)$. On the other hand, if ξ is not separable over K, then Lemma 9 shows that, for every $\lambda \in L$, there is a unique Δ in $\mathrm{Der}_K(K(\xi), L)$ such that $\Delta\xi = \lambda$.

In the next theorem, K^* denotes a field between K and L, and $\xi_1, \xi_2, \ldots, \xi_n$ belong to L.

Theorem 7. _Suppose that_ $\xi_1, \xi_2, \ldots, \xi_n$ _are all algebraic and separable over_ K^*. _Then each_ D _in_ $\mathrm{Der}_K(K^*, L)$ _has a unique extension to a member of_ $\mathrm{Der}_K(K^*(\xi_1, \ldots, \xi_n), L)$.

This follows from successive applications of Lemma 9 Cor. 1.

Theorem 8. _Let_ $\xi_1, \xi_2, \ldots, \xi_n$ _belong to_ L _and be algebraic over_ K. _Then_

$$\mathrm{Der}_K(K(\xi_1, \ldots, \xi_n), L) = 0$$

if and only if $K(\xi_1, \xi_2, \ldots, \xi_n)$ _is a separable extension of_ K.

Proof. If $K(\xi_1, \xi_2, \ldots, \xi_n)$ is separable over K, then $\mathrm{Der}_K(K(\xi_1, \ldots, \xi_n), L) = 0$ by virtue of Theorem 7. Now suppose that $K(\xi_1, \xi_2, \ldots, \xi_n)$ is not separable over K. Then K has characteristic p, where p is a prime. Let K' be the separable closure of K in $K(\xi_1, \xi_2, \ldots, \xi_n)$. Then $K' \neq K(\xi_1, \xi_2, \ldots, \xi_n)$ and there exists a positive integer ν such that $\xi_i^{p^\nu} \in K'$ for $i = 1, 2, \ldots, n$.

Let j be the smallest integer such that

$$K'(\xi_1, \xi_2, \ldots, \xi_j) = K(\xi_1, \xi_2, \ldots, \xi_n)$$

and put $K^* = K'(\xi_1, \xi_2, \ldots, \xi_{j-1})$. Then $K^*(\xi_j) = K(\xi_1, \xi_2, \ldots, \xi_n)$ and ξ_j is not separable over K^*. Accordingly

$$\mathrm{Der}_{K^*}(K(\xi_1, \ldots, \xi_n), L) = \mathrm{Der}_{K^*}(K^*(\xi_j), L)$$

is not zero by Lemma 9 Cor. 2. It follows _a fortiori_ that $\mathrm{Der}_K(K(\xi_1, \ldots, \xi_n), L) \neq 0$.

Suppose that $\xi_1, \xi_2, \ldots, \xi_n$ belong to L but are not necessarily algebraic over K. Let the transcendence degree of $K(\xi_1, \xi_2, \ldots, \xi_n)$ over K be r.

Definition. We say that $K(\xi_1, \xi_2, \ldots, \xi_n)$ is a 'separable' extension of K if there is a transcendence base $\eta_1, \eta_2, \ldots, \eta_r$, for $K(\xi_1, \xi_2, \ldots, \xi_n)$ over K, such that $K(\xi_1, \xi_2, \ldots, \xi_n)$ is a separable algebraic extension of $K(\eta_1, \eta_2, \ldots, \eta_r)$.

Such a transcendence base is called a <u>separating</u> transcendence base. Thus $K(\xi_1, \xi_2, \ldots, \xi_n)$ is separable over K if and only if a separating transcendence base exists. Of course if K has characteristic zero, then $K(\xi_1, \xi_2, \ldots, \xi_n)$ is necessarily a separable extension of K.

Lemma 10.† *Let $\xi_1, \xi_2, \ldots, \xi_n$ belong to L and let $K(\xi_1, \xi_2, \ldots, \xi_n)$ be a separable extension of K. Then the dimension of $\mathrm{Der}_K(K(\xi_1, \ldots, \xi_n), L)$, considered as a vector space over L, is equal to the transcendence degree of $K(\xi_1, \xi_2, \ldots, \xi_n)$ over K.*

Proof. Let $\eta_1, \eta_2, \ldots, \eta_r$ be a separating transcendence base. By Example 4 in section (4.2), the r derivations

$$\frac{\partial}{\partial \eta_i} : K(\eta_1, \eta_2, \ldots, \eta_r) \to L$$

form a base for the L-space $\mathrm{Der}_K(K(\eta_1, \ldots, \eta_r), L)$ and, by Theorem 7 $\partial/\partial\eta_i$ has a unique extension, D_i say, to a derivation of $K(\xi_1, \xi_2, \ldots, \xi_n)$ into L. It is clear that each member of $\mathrm{Der}_K(K(\xi_1, \ldots, \xi_n), L)$ has a unique representation in the form

$$\lambda_1 D_1 + \lambda_2 D_2 + \ldots + \lambda_r D_r,$$

where $\lambda_i \in L$. The theorem follows.

Theorem 9. *Suppose that $\xi_1, \xi_2, \ldots, \xi_n$ belong to L and that $\mathrm{Der}_K(K(\xi_1, \ldots, \xi_n), L) = 0$. Then each ξ_i is separable and algebraic over K.*

Proof. By Theorem 8, it is enough to prove that the ξ_i are algebraic over K. Assume the contrary and let $\eta_1, \eta_2, \ldots, \eta_r$ be a transcendence base for $K(\xi_1, \xi_2, \ldots, \xi_n)$ over K. Then $r \geq 1$ and

† The converse is also true. See Theorem 12.

each ξ_i is algebraic over $K(\eta_1, \eta_2, \ldots, \eta_r)$. Now $K(\xi_1, \xi_2, \ldots, \xi_n)$ cannot be separable over $K(\eta_1, \eta_2, \ldots, \eta_r) = K^*$ (say) for otherwise Lemma 10 gives a contradiction. Hence, by Theorem 8,

$$\mathrm{Der}_{K^*}(K(\xi_1, \ldots, \xi_n), L) = \mathrm{Der}_{K^*}(K^*(\xi_1, \ldots, \xi_n), L)$$

is non-zero and therefore *a fortiori*

$$\mathrm{Der}_K(K(\xi_1, \ldots, \xi_n), L) \neq 0.$$

Thus in any event we arrive at a contradiction.

Theorem 10. *Let $\xi_1, \xi_2, \ldots, \xi_n$ belong to L and let $F_j(X_1, X_2, \ldots, X_n)$, where $1 \leq j \leq n$, belong to $K[X_1, X_2, \ldots, X_n]$. If now $F_j(\xi_1, \xi_2, \ldots, \xi_n) = 0$ for all j and the determinant $|\partial F_j/\partial \xi_i|$ is not zero, then $K(\xi_1, \xi_2, \ldots, \xi_n)$ is a separable algebraic extension of K.*

Proof. By Theorem 9, it will suffice to show that $\mathrm{Der}_K(K(\xi_1, \ldots, \xi_n), L) = 0$. Suppose then that D is a derivation of $K(\xi_1, \xi_2, \ldots, \xi_n)$ into L over K. The theorem will follow if we show that $D\xi_i = 0$ for all i. However

$$\sum_{i=1}^{n} \frac{\partial F_j}{\partial \xi_i} D\xi_i = 0$$

for $j = 1, 2, \ldots, n$ and therefore $D\xi_i = 0$ as required.

Once again let $\xi_1, \xi_2, \ldots, \xi_n$ belong to L. The polynomials $F(X_1, X_2, \ldots, X_n)$, with coefficients in K, that satisfy $F(\xi_1, \xi_2, \ldots, \xi_n) = 0$ form a prime ideal, $\mathfrak{A}$ say, in $K[X_1, X_2, \ldots, X_n]$. Let the polynomials $F_j(X_1, X_2, \ldots, X_n)$, where $1 \leq j \leq m$, generate $\mathfrak{A}$.

Theorem 11. *Let the notation be as explained above. Further let the rank of the matrix $\|\partial F_j/\partial \xi_i\|$, where $1 \leq i \leq n$ and $1 \leq j \leq m$, be $n - s$ and let the transcendence degree of $K(\xi_1, \xi_2, \ldots, \xi_n)$ over K be r. Then $r \leq s$. Moreover $r = s$ if and only if $K(\xi_1, \xi_2, \ldots, \xi_n)$ is separable over K.*

Proof. By Lemma 6 Cor., $\mathrm{Der}_K(K(\xi_1, \ldots, \xi_n), L)$ and $\mathrm{Der}_K(K[\xi_1, \xi_2, \ldots, \xi_n], L)$ are isomorphic L-spaces. On the other hand Theorem 6 shows that $\mathrm{Der}_K(K[\xi_1, \ldots, \xi_n], L)$ is isomorphic to the space of solutions (in L) of the equations

$$\frac{\partial F_j}{\partial \xi_1}\lambda_1 + \frac{\partial F_j}{\partial \xi_2}\lambda_2 + \ldots + \frac{\partial F_j}{\partial \xi_n}\lambda_n = 0 \quad (1 \le j \le m).$$

Accordingly s _is the dimension of_ $\mathrm{Der}_K(K(\xi_1, \ldots, \xi_n), L)$ _considered as a vector space over_ L. Hence, by Lemma 10, if $K(\xi_1, \xi_2, \ldots, \xi_n)$ is a separable extension of K, then $r = s$.

Now let us renumber $F_1, F_2, \ldots, F_m$ and $\xi_1, \xi_2, \ldots, \xi_n$ so that the $(n - s) \times (n - s)$ matrix $\| \partial F_\nu / \partial \xi_\mu \|$, where $1 \le \nu \le n-s$ and $s+1 \le \mu \le n$, has a non-zero determinant.† Then Theorem 10 shows that $K(\xi_1, \xi_2, \ldots, \xi_n)$ is a separable algebraic extension of $K(\xi_1, \xi_2, \ldots, \xi_s)$. Thus, in particular, $r \le s$. Moreover if $r = s$, then $\xi_1, \xi_2, \ldots, \xi_r$ must be a separating transcendence base for $K(\xi_1, \xi_2, \ldots, \xi_n)$ over K and therefore the extension is separable.

Theorem 12. _Let_ $\xi_1, \xi_2, \ldots, \xi_n$ _belong to_ L, _let_ $K(\xi_1, \xi_2, \ldots, \xi_n)$ _have transcendence degree_ r _over_ K, _and let_ $\mathrm{Der}_K(K(\xi_1, \ldots, \xi_n), L)$ _be an_ L-_space of dimension_ s. _Then_ $r \le s$ _and there is equality if and only if_ $K(\xi_1, \xi_2, \ldots, \xi_n)$ _is a separable extension of_ K.

All these assertions were established during the proof of the last theorem. At the same time we also proved

Theorem 13. _Let_ $\xi_1, \xi_2, \ldots, \xi_n$ _belong to_ L _and suppose that_ $K(\xi_1, \xi_2, \ldots, \xi_n)$ _is separable over_ K. _Then a separating transcendence base can be chosen from among_ $\xi_1, \xi_2, \ldots, \xi_n$.

The field K can only have non-separable extensions if its characteristic is a prime p. Suppose that this is the case and let us regard L as being contained in its algebraic closure $\bar{L}$. Put

$$K^{1/p} = \{\alpha \mid \alpha \in \bar{L} \text{ and } \alpha^p \in K\}.$$

† The subsequent argument is trivial if $s = n$.

This is a subfield of $\bar{L}$ containing K.

Lemma 11. Suppose that $\xi_1, \xi_2, \ldots, \xi_n$ belong to L and are algebraic and separable over K. Then $K(\xi_1, \xi_2, \ldots, \xi_n)$ and $K^{1/p}$ are linearly disjoint over K.

Proof. Assume the contrary. Then we can find $\eta_1, \eta_2, \ldots, \eta_m$, in $K(\xi_1, \xi_2, \ldots, \xi_n)$, linearly independent over K but not linearly independent over $K^{1/p}$. Let us arrange that m is as small as possible. Then there exist non-zero elements $\gamma_1, \gamma_2, \ldots, \gamma_m$, in $K^{1/p}$, such that $\gamma_1\eta_1 + \gamma_2\eta_2 + \ldots + \gamma_m\eta_m = 0$ and we can fix it that $\gamma_m = 1$. Note that $\gamma_1, \gamma_2, \ldots, \gamma_m$ are not all in K and therefore $K(\gamma_1, \gamma_2, \ldots, \gamma_m)$ is not a separable extension of K.

By Theorem 8, $\mathrm{Der}_K(K(\gamma_1, \ldots, \gamma_m), K^{1/p})$ contains a non-zero derivation D (say). Thus $D\gamma_i \in K^{1/p}$ and at least one of $D\gamma_1, D\gamma_2, \ldots, D\gamma_{m-1}$ is not zero. Naturally $D\gamma_m = 0$.

Next $K(\gamma_1, \ldots, \gamma_m, \xi_1, \ldots, \xi_n)$ is a separable algebraic extension of $K(\gamma_1, \ldots, \gamma_m)$ and therefore, by Theorem 7, D has an extension which belongs to

$$\mathrm{Der}_K(K(\gamma_1, \ldots, \gamma_m, \xi_1, \ldots, \xi_n), \bar{L}),$$

where $\bar{L}$ denotes the algebraic closure of L. (This extension will also be denoted by D.) Now Theorem 8 shows that D must vanish on $K(\xi_1, \xi_2, \ldots, \xi_n)$ and, in particular, $D\eta_i = 0$ for $i = 1, 2, \ldots, m$. Accordingly

$$\sum_{i=1}^{m-1} (D\gamma_i)\eta_i = 0$$

and this contradicts the minimal property of the integer m.

Lemma 12. Suppose that the characteristic of K is a prime p and let $\xi_1, \xi_2, \ldots, \xi_n$ belong to L. If now $K(\xi_1, \xi_2, \ldots, \xi_n)$ is a separable extension of K, then $K(\xi_1, \xi_2, \ldots, \xi_n)$ and $K^{1/p}$ are linearly disjoint over K.

Proof. Let $\gamma_1, \gamma_2, \ldots, \gamma_m$ belong to $K^{1/p}$ and be linearly independent over K. It will suffice to prove that they are linearly independent over $K(\xi_1, \xi_2, \ldots, \xi_n)$.

Take a separating transcendence base for $K(\xi_1, \xi_2, \ldots, \xi_n)$ over K, say $\eta_1, \eta_2, \ldots, \eta_r$, and put $K_1 = K(\eta_1, \eta_2, \ldots, \eta_r)$. Then $\gamma_1, \gamma_2, \ldots, \gamma_m$ belong to $K_1^{1/p}$ and it is clear that they are linearly independent over K_1. That $\gamma_1, \gamma_2, \ldots, \gamma_m$ are linearly independent over $K(\xi_1, \xi_2, \ldots, \xi_n) = K_1(\xi_1, \xi_2, \ldots, \xi_n)$ now follows from Lemma 11.

The lemma just proved has a converse. The full result is contained in

Theorem 14. Let the characteristic of K be the prime p and let $\xi_1, \xi_2, \ldots, \xi_n$ belong to L. Then the following statements are equivalent:

(a) $K(\xi_1, \xi_2, \ldots, \xi_n)$ is a separable extension of K;

(b) $K(\xi_1, \xi_2, \ldots, \xi_n)$ and $K^{1/p}$ are linearly disjoint over K.

Proof. We shall assume (b) and show that (a) follows. (In view of Lemma 12 this will be sufficient.) The demonstration uses induction on n and we begin by observing that for $n = 0$ the result in question is obvious. From here on it will be assumed that $n \geq 1$ and that we know that (b) implies (a) in the case of extensions generated by only $n - 1$ elements. Note that we may suppose that $\xi_1, \xi_2, \ldots, \xi_n$ are not algebraically independent over K for otherwise there would be no problem.

There exists a non-null polynomial, $F(X_1, X_2, \ldots, X_n)$ say, with coefficients in K such that $F(\xi_1, \xi_2, \ldots, \xi_n) = 0$. In what follows $F(X_1, X_2, \ldots, X_n)$ is to be chosen so that its degree is as small as possible. Put

$$F_i(X_1, X_2, \ldots, X_n) = \frac{\partial}{\partial X_i} F(X_1, X_2, \ldots, X_n).$$

We claim that it will suffice to show that at least one of the $F_i(X_1, X_2, \ldots, X_n)$ is not null. For suppose that $F_j(X_1, X_2, \ldots, X_n)$ is not null. Then $F_j(\xi_1, \xi_2, \ldots, \xi_n) \neq 0$ and therefore ξ_j is separably algebraic over $K(\xi_1, \ldots, \xi_{j-1}, \xi_{j+1}, \ldots, \xi_n)$. But induction shows that

$K(\xi_1, \ldots, \xi_{j-1}, \xi_{j+1}, \ldots, \xi_n)$ is separable over K and, by combining these facts, we conclude that $K(\xi_1, \xi_2, \ldots, \xi_n)$ is separable over K.

From this point onwards we assume that, for each i, $F_i(X_1, X_2, \ldots, X_n)$ is the null polynomial and we seek a contradiction. Our new assumption ensures that

$$F(X_1, X_2, \ldots, X_n) = [G(X_1, X_2, \ldots, X_n)]^p$$

for some $G(X_1, X_2, \ldots, X_n)$ in $K^{1/p}[X_1, X_2, \ldots, X_n]$. Choose $\gamma_1, \gamma_2, \ldots, \gamma_s$ in $K^{1/p}$ so that they are linearly independent over K and

$$G(X_1, X_2, \ldots, X_n) = \sum_{\nu=1}^{s} \gamma_\nu G_\nu(X_1, X_2, \ldots, X_n),$$

where $G_\nu(X_1, X_2, \ldots, X_n) \in K[X_1, X_2, \ldots, X_n]$. We can choose μ so that $G(X_1, X_2, \ldots, X_n)$ and $G_\mu(X_1, X_2, \ldots, X_n)$ have the same degree and this will be smaller than the degree of $F(X_1, X_2, \ldots, X_n)$. Consequently $G_\mu(\xi_1, \xi_2, \ldots, \xi_n) \neq 0$. On the other hand $G(\xi_1, \xi_2, \ldots, \xi_n)=0$, because $F(\xi_1, \xi_2, \ldots, \xi_n) = 0$, and $\gamma_1, \gamma_2, \ldots, \gamma_s$ are linearly independent over $K(\xi_1, \xi_2, \ldots, \xi_n)$. It follows that $G_\mu(\xi_1, \xi_2, \ldots, \xi_n) = 0$ and with this we have the desired contradiction.

Corollary. Let $\xi_1, \xi_2, \ldots, \xi_n$ belong to L and let $\eta_1, \eta_2, \ldots, \eta_m$ belong to $K(\xi_1, \xi_2, \ldots, \xi_n)$. If now $K(\xi_1, \xi_2, \ldots, \xi_n)$ is a separable extension of K, then $K(\eta_1, \eta_2, \ldots, \eta_m)$ is also a separable extension of K.

This holds regardless of the value of the characteristic of K.

We recall that $K(\xi_1, \xi_2, \ldots, \xi_n)$ is called a regular extension of K if $K(\xi_1, \xi_2, \ldots, \xi_n)$ and the algebraic closure of K are linearly disjoint over K.

Theorem 15. Let $\xi_1, \xi_2, \ldots, \xi_n$ belong to L and suppose that $K(\xi_1, \xi_2, \ldots, \xi_n)$ is a regular extension of K. Then $K(\xi_1, \xi_2, \ldots, \xi_n)$ is a separable extension of K.

Proof. If the characteristic of K is zero, then the assertion is obvious, whereas if the characteristic is a prime p, then Theorem 15 follows from Theorem 14.

It may be shown† that $K(\xi_1, \xi_2, \ldots, \xi_n)$ is a regular extension of K if and only if (i) $K(\xi_1, \xi_2, \ldots, \xi_n)$ is separable over K, and (ii) K is algebraically closed in $K(\xi_1, \xi_2, \ldots, \xi_n)$. However we shall not be using this result.

4.4 Tangent spaces and simple points

It is time to apply the results of the preceding sections to the theory of affine sets. Accordingly, throughout section (4.4), V will denote an affine set (defined over K) and L will denote an extension field of K. When $x \in V$ we use M_x to describe the corresponding rational maximal ideal of $K[V]$ and, as before, $Q_{V,x}$ denotes the local ring of V at x. On this occasion the unique maximal ideal $M_x Q_{V,x}$, of $Q_{V,x}$ will be designated by $M_{V,x}$.

From time to time some auxiliary notation will be needed. To avoid tedious repetition we shall explain it here once for all.

The notation arises in the following way. We choose $\xi_1, \xi_2, \ldots, \xi_n$ in $K[V]$ so that

$$K[V] = K[\xi_1, \xi_2, \ldots, \xi_n]. \tag{4.4.1}$$

(There are many ways of doing this.) Then we let $X_1, X_2, \ldots, X_n$ be indeterminates and construct the surjective homomorphism

$$\theta : K[X_1, X_2, \ldots, X_n] \to K[\xi_1, \xi_2, \ldots, \xi_n] \tag{4.4.2}$$

of K-algebras in which $\theta(X_i) = \xi_i$. Next we put

$$\mathfrak{A} = \operatorname{Ker} \theta \tag{4.4.3}$$

and select a finite set of polynomials, say

$$F_j(X_1, X_2, \ldots, X_n) \text{ where } 1 \le j \le m,$$

---

† See [(11) Theorem 5, p. 18].

which generates $\mathfrak{A}$. Finally when $x \in V$ we write

$$x_i = \xi_i(x) \tag{4.4.4}$$

for $1 \leq i \leq m$. Naturally our main interest will be in results whose statements do not involve $\xi_1, \xi_2, \ldots, \xi_n$ nor any of the entities defined in terms of them.

Let $x \in V$. This point gives rise to a homomorphism

$$\omega_x : K[V] \to K \tag{4.4.5}$$

in which $\omega_x(f) = f(x)$. Put

$$\mathrm{Der}_K(V, x) = \mathrm{Der}_K(K[V], K, \omega_x). \tag{4.4.6}$$

The members of $\mathrm{Der}_K(V, x)$ are called <u>local derivations</u> on V at x. Thus if $D \in \mathrm{Der}_K(V, x)$, then $D : K[V] \to K$ is K-linear and

$$D(fg) = (Df)g(x) + f(x)(Dg)$$

for all f, g in $K[V]$.

Definition. Let $x \in V$. Then the K-space $\mathrm{Der}_K(V, x)$ is called the 'tangent space' to V at x.

Of course from elementary geometry one has an intuitive idea of a tangent space. It is therefore desirable to show how the above definition is connected with our previously acquired concept. This, however, will be postponed for the moment. Our immediate concern will be to examine the effect of K-morphisms on tangent spaces.

To this end suppose that $\phi : V \to W$ is a K-morphism of affine sets and let $\phi^* : K[W] \to K[V]$ be the corresponding homomorphism of K-algebras. If $x \in V$ and ω_x is defined as in (4.4.5), then, by (4.1.6), we have a K-linear mapping

$$\mathrm{Der}_K(K[V], K, \omega_x) \to \mathrm{Der}_K(K[W], K, \omega_x \circ \phi^*).$$

But $\omega_x \circ \phi^* : K[W] \to K$ maps g into $g(\phi(x))$. Thus we obtain

Theorem 16. *Let $\phi : V \to W$ be a K-morphism of affine sets and let $x \in V$. Then there is a linear mapping*

$$d(\phi, x) : \mathrm{Der}_K(V, x) \to \mathrm{Der}_K(W, \phi(x))$$

of K-spaces in which $D \in \mathrm{Der}_K(V, x)$ is mapped into $D \circ \phi^*$.

Obviously if ϕ is a K-isomorphism, then $d(\phi, x)$ is an isomorphis of K-spaces. Also if $\psi : W \to U$ is a further K-morphism, then

$$d(\psi, \phi(x)) \circ d(\phi, x) = d(\psi \circ \phi, x). \tag{4.4.7}$$

Let $x \in V$ and consider the equations

$$\begin{aligned} \frac{\partial F_1}{\partial x_1} a_1 + \frac{\partial F_1}{\partial x_2} a_2 + \dots + \frac{\partial F_1}{\partial x_n} a_n &= 0 \\ \frac{\partial F_2}{\partial x_1} a_1 + \frac{\partial F_2}{\partial x_2} a_2 + \dots + \frac{\partial F_2}{\partial x_n} a_n &= 0 \\ \cdot \qquad \cdot \qquad \cdots \qquad \cdot & \\ \cdot \qquad \cdot \qquad \cdots \qquad \cdot & \\ \frac{\partial F_m}{\partial x_1} a_1 + \frac{\partial F_m}{\partial x_2} a_2 + \dots + \frac{\partial F_m}{\partial x_n} a_n &= 0. \end{aligned} \tag{4.4.8}$$

We regard these as equations to be solved, in K, for $a_1, a_2, \dots, a_n$. Theorems 5 and 6 applied to the present situation now give

Theorem 17. *There is a natural isomorphism between the K-space formed by the solutions of the equations (4.4.8) and the tangent space, $\mathrm{Der}_K(V, x)$, of V at x.*

Corollary. *Let $n - s$ be the rank of the matrix $\| \partial F_j / \partial x_i \|$. Then the dimension of $\mathrm{Der}_K(V, x)$, considered as a vector space over K, is s*

We shall now digress briefly in order to explain the name *tangent space* that has been given to $\mathrm{Der}_K(V, x)$.

Suppose, for the moment, that V is a closed subset of K^n. Each indeterminate X_i induces a coordinate function on V and if we call these functions $\xi_1, \xi_2, \dots, \xi_n$, then (4.4.1) holds. The ideal $\mathfrak{A}$ of (4.4.3) nc consists of all polynomials $F(X_1, X_2, \dots, X_n)$ which vanish everywhere on V. Moreover, if $x \in V$, then $x_1, x_2, \ldots, x_n$, as defined by (4.4.4), are just the coordinates of x considered as a point of K^n.

Let $x \in V$ and suppose that $a_1, a_2, \ldots, a_n$ belong to K. Further let Y be a new indeterminate. If $F(X_1, X_2, \ldots, X_n)$ belongs to $\mathfrak{A}$, then the constant term in

$$F(x_1 + a_1 Y, x_2 + a_2 Y, \ldots, x_n + a_n Y)$$

is zero and the coefficient of Y is

$$\sum_{i=1}^{n} \frac{\partial F}{\partial x_i} a_i .$$

If this coefficient is zero for all $F \in \mathfrak{A}$, then $(a_1, a_2, \ldots, a_n)$ is called a _tangent vector_ to V at x. The tangent vectors to V at x form a vector space, $T_{V,x}$ say, over K. Evidently $(a_1, a_2, \ldots, a_n)$ is in $T_{V,x}$ if and only if the equations (4.4.8) are satisfied. We can therefore restate Theorem 17, for this special situation, as

Theorem 18. _Let_ V _be a closed subset of_ K^n _and let_ $x \in V$. _Then the space_ $T_{V,x}$ _of tangent vectors and the abstract tangent space_ $\mathrm{Der}_K(V, x)$ _are naturally isomorphic K-spaces._

This ends the digression. From now on we assume, once again, that V is an abstract affine set.

Let V be _irreducible_ and suppose that $x \in V$. If now $D \in \mathrm{Der}_K(V, x)$ then, since

$$Q_{V,x} = (K[V])_{M_x},$$

our discussion of derivations and fractions shows that there is a unique $\overline{D} \in \mathrm{Der}_K(Q_{V,x}, K)$ which extends D. (Note that, because there is only one possible homomorphism $Q_{V,x} \to K$ of K-algebras, there is no need to make the notation more explicit.)

Theorem 19. _Let_ V _be an irreducible affine set and let_ $x \in V$. _Then each_ $D \in \mathrm{Der}_K(V, x)$ _has a unique extension_ $\overline{D}$ _in_ $\mathrm{Der}_K(Q_{V,x}, K)$ _and the mapping_

$$\mathrm{Der}_K(V, x) \to \mathrm{Der}_K(Q_{V,x}, K)$$

<u>in which</u> $D \mapsto \overline{D}$ <u>is an isomorphism of</u> K-<u>spaces.</u>

This follows from Lemma 6 Cor.

The next theorem provides another K-space that is isomorphic to $\mathrm{Der}_K(Q_{V,x}, K)$. This result does not require V to be irreducible. We recall that $M_{V,x}$ is being used to denote the maximal ideal of $Q_{V,x}$.

Suppose that $D \in \mathrm{Der}_K(Q_{V,x}, K)$. By restriction D gives rise to a K-linear mapping $M_{V,x} \to K$ and this vanishes on $M_{V,x}^2$. Thus D induces a K-linear mapping

$$\overline{\Delta} : M_{V,x}/M_{V,x}^2 \to K,$$

i.e. $\overline{\Delta} \in \mathrm{Hom}_K(M_{V,x}/M_{V,x}^2, K)$. Furthermore the mapping

$$\mathrm{Der}_K(Q_{V,x}, K) \to \mathrm{Hom}_K(M_{V,x}/M_{V,x}^2, K) \tag{4.4.9}$$

which sends D into $\overline{\Delta}$ is K-linear.

Theorem 20. <u>Let</u> x <u>be a point of the affine set</u> V. <u>Then (4.4.9) is an isomorphism of the</u> K-<u>space</u> $\mathrm{Der}_K(Q_{V,x}, K)$ <u>on to the</u> K-<u>space</u> $\mathrm{Hom}_K(M_{V,x}/M_{V,x}^2, K)$.

Proof. Suppose that $\xi \in Q_{V,x}$. Then ξ has a unique representation in the form $\xi = k + u$, where $k \in K$ and $u \in M_{V,x}$. Also if $D \in \mathrm{Der}_K(Q_{V,x}, K)$, then $D\xi = Du$. Hence if D belongs to the kernel of (4.4.9), then $Du = 0$ and therefore $D\xi = 0$. Thus our mapping is an injection.

Next assume that

$$\overline{\Delta} : M_{V,x}/M_{V,x}^2 \to K$$

is K-linear and let $\Delta : M_{V,x} \to K$ be obtained by combining $\overline{\Delta}$ with the natural mapping

$$M_{V,x} \to M_{V,x}/M_{V,x}^2.$$

Now define $D : Q_{V,x} \to K$ as follows. If $\xi \in Q_{V,x}$ and $\xi = k + u$

(where $k \in K$ and $u \in M_{V,x}$) put $D\xi = \Delta u$. An easy verification shows that $D \in \mathrm{Der}_K(Q_{V,x}, K)$ and it is clear that $D \mapsto \bar{\Delta}$ under (4.4.9).

The next series of results is aimed at obtaining information about the actual dimension of the tangent space $\mathrm{Der}_K(V, x)$.

Theorem 21. *Let* V *be an irreducible affine set defined over* K. *Then* $K(V)$ *is a separable extension of* K.

This follows by combining Theorem 29 of Chapter 3 with Theorem 15 of this chapter.

Corollary. *Let* V *be an irreducible affine set and put* $\mathrm{Dim}\, V = r$. *Then (with the notation explained at the beginning of this section) the matrix* $\| \partial F_j / \partial \xi_i \|$ *has rank* $n - r$.

Proof. We have $K(V) = K(\xi_1, \xi_2, \ldots, \xi_n)$ and we have just shown that this is a separable extension of K. The corollary therefore follows from Theorem 11.

Still supposing that V is irreducible, let $x \in V$. If ω_x is the homomorphism described in (4.4.5), then

$$\omega_x \left(\frac{\partial F_j}{\partial \xi_i} \right) = \frac{\partial F_j}{\partial x_i}$$

It therefore follows, from the last corollary, that the rank of $\| \partial F_j / \partial x_i \|$ is at most $n - r$. This observation, combined with Theorem 17 Cor., yields

Theorem 22. *Let* x *be a point of the irreducible variety* V. *Then the dimension of the tangent space to* V *at* x *is at least* $\mathrm{Dim}\, V$.

We are now ready for the

Definition. A point x of the irreducible affine set V is called a 'simple point' if the dimension of the tangent space $\mathrm{Der}_K(V, x)$, considered as a vector space over K, is equal to $\mathrm{Dim}\, V$. Points of V which are not simple are called 'multiple points'.

Theorem 23. Let V be an irreducible affine set and suppose that $x \in V$. Then x is simple on V if and only if the dimension of the K-space $M_{V,x}/M_{V,x}^2$ is equal to Dim V.

Proof. By Theorems 19 and 20, x is a simple point of V if and only if the dimension of the K-space

$$\mathrm{Hom}_K(M_{V,x}/M_{V,x}^2, K)$$

equals Dim V. However this happens when and only when $M_{V,x}/M_{V,x}^2$ is a K-space whose dimension is Dim V.

Theorem 24. Let V be an irreducible affine set. Then the simple points of V form a non-empty open subset.

Remark. Note that this result shows that an irreducible affine set has at least one simple point.

Proof. Put $r = \mathrm{Dim}\, V$ and consider the matrix $\| \partial F_j / \partial X_i \|$, where the notation is that introduced at the beginning of the section. The subdeterminants† of order $n - r$ will be certain polynomials, say $G_j(X_1, X_2, \ldots, X_n)$ where $1 \leq j \leq s$, and, by Theorem 21 Cor., we have $G_j(\xi_1, \xi_2, \ldots, \xi_n) \neq 0$ for at least one value of j. Without loss of generality we may suppose that $G_j(\xi_1, \xi_2, \ldots, \xi_n)$ is non-zero for every j in the range $1 \leq j \leq \mu$, whereas it is zero whenever $\mu+1 \leq j \leq s$. Note that $\mu \geq 1$.

Let $x \in V$ and define $x_1, x_2, \ldots, x_n$ as in (4.4.4). We obtain $\partial F_j / \partial x_i$ by applying the homomorphism ω_x, of (4.4.5), to $\partial F_j / \partial \xi_i$. Since $\| \partial F_j / \partial \xi_i \|$ has rank $n - r$, the rank of $\| \partial F_j / \partial x_i \|$ is at most $n - r$; and x is a simple point precisely when the latter rank has this value. Thus to sum up: x is simple on V if and only if there is a j $(1 \leq j \leq \mu)$ such that $G_j(x_1, x_2, \ldots, x_n) \neq 0$.

Suppose that $1 \leq j \leq \mu$ and put $f_j = G_j(\xi_1, \xi_2, \ldots, \xi_n)$. Then $f_j \in K[V]$, $f_j \neq 0$, and $f_j(x) = G_j(x_1, x_2, \ldots, x_n)$. Accordingly

† Note that if $r = n$, then all the points of V are simple.

$$N_j = \{x \mid x \in V \text{ and } f_j(x) \neq 0\}$$

is a non-empty open subset of V and each of its points is simple on V. Finally $N_1 \cup N_2 \cup \ldots \cup N_\mu$ is the set of all simple points.

Theorem 25. Let $\phi : V \to W$ be a K-isomorphism of affine sets and suppose that V, W are irreducible. If now $x \in V$, then x is simple on V if and only if $\phi(x)$ is simple on W.

Proof. Since Dim V = Dim W, this follows from the remarks that come immediately after Theorem 16.

We next make a brief investigation of the effect of enlarging the ground field on tangent spaces and simple points. We recall that L denotes an extension field of K.

Let $f \in K[V]$. Then f has a natural prolongation, $\hat{f}$ say, to a function on $V^{(L)}$. Since $K[V] = K[\xi_1, \xi_2, \ldots, \xi_n]$, we have

$$L[V^{(L)}] = L[\hat{\xi}_1, \hat{\xi}_2, \ldots, \hat{\xi}_n]. \tag{4.4.10}$$

Denote by

$$\hat{\theta} : L[X_1, X_2, \ldots, X_n] \to L[\hat{\xi}_1, \hat{\xi}_2, \ldots, \hat{\xi}_n] \tag{4.4.11}$$

the surjective homomorphism of L-algebras in which $\hat{\theta}(X_i) = \hat{\xi}_i$, and put

$$\mathfrak{A} = \text{Ker } \hat{\theta}. \tag{4.4.12}$$

Let $G(X_1, X_2, \ldots, X_n)$ belong to $L[X_1, X_2, \ldots, X_n]$ and write G in the form

$$G = \lambda_1 G_1 + \lambda_2 G_2 + \ldots + \lambda_q G_q,$$

where $\lambda_1, \lambda_2, \ldots, \lambda_q$ belong to L and are linearly independent over K and $G_j = G_j(X_1, X_2, \ldots, X_n)$ belongs to $K[X_1, X_2, \ldots, X_n]$. Then

$$\hat{\theta}(G) = \sum_{i=1}^{q} \lambda_i \widehat{\theta(G_i)}$$

and now it follows that $G \in \hat{\mathfrak{A}}$ if and only if each G_i is in $\mathfrak{A}$ (see (4.4.3)). Consequently

$$\hat{\mathfrak{A}} = \mathfrak{A}L[X_1, X_2, \ldots, X_n]$$

and therefore the m polynomials $F_j(X_1, X_2, \ldots, X_n)$ which generate $\mathfrak{A}$ in $K[X_1, X_2, \ldots, X_n]$ also generate $\hat{\mathfrak{A}}$ in $L[X_1, X_2, \ldots, X_n]$.

Lemma 13. _Let_ V _be an affine set defined over_ K, _let_ $x \in V$, _and let_ L _be an extension field of_ K. _Then the dimension of the_ K-_space_ $\mathrm{Der}_K(V, x)$ _is equal to the dimension of the_ L-_space_ $\mathrm{Der}_L(V^{(L)}, x)$.

Proof. Since $\hat{\xi}_i(x) = \xi_i(x) = x_i$, Theorem 17 Cor. shows that each of the spaces has dimension s, where n - s is the rank of the matrix $\| \partial F_j / \partial x_i \|$.

Once again suppose that $x \in V \subseteq V^{(L)}$. If

$$\omega_x : K[V] \to K$$

is the homomorphism in which $\omega_x(f) = f(x)$, then, since $K[V]^L = L[V^{(L)}]$ and $K^L = L$,

$$\omega_x^L : L[V^{(L)}] \to L$$

is the homomorphism of L-algebras in which $\omega_x^L(u) = u(x)$. Let $D \in \mathrm{Der}_K(K[V], K, \omega_x)$. Then, by Lemma 7,

$$D^L \in \mathrm{Der}_L(L[V^{(L)}], L, \omega_x^L).$$

Thus when D belongs to $\mathrm{Der}_K(V, x)$, D^L is in $\mathrm{Der}_L(V^{(L)}, x)$.

Theorem 26. _Let_ V _be an affine set defined over_ K, _let_ $x \in V$, _and let_ L _be an extension field of_ K. _If now_ $D_1, D_2, \ldots, D_q$ _is a base for the_ K-_space_ $\mathrm{Der}_K(V, x)$, _then_ $D_1^L, D_2^L, \ldots, D_q^L$ _is a base for the_ L-_space_ $\mathrm{Der}_L(V^{(L)}, x)$.

Proof. By Lemma 13, it will suffice to show that the derivations D_i^L are linearly independent over L. Now the rows of the matrix

$$\begin{Vmatrix} D_1\xi_1 & D_1\xi_2 \cdots & D_1\xi_n \\ D_2\xi_1 & D_2\xi_2 & D_2\xi_n \\ \cdot & \cdot \;\cdots & \cdot \\ \cdot & \cdot \;\cdots & \cdot \\ D_q\xi_1 & D_q\xi_2 & D_q\xi_n \end{Vmatrix}$$

are linearly independent over K because otherwise $D_1, D_2, \ldots, D_q$ would not be linearly independent over this field. Next† $D_i^L\hat{\xi}_j = D_i\xi_j$ (by the definition of D_i^L) and therefore the rows of the matrix

$$\|D_i^L\hat{\xi}_j\| \quad (1 \le i \le q,\ 1 \le j \le n)$$

must be linearly independent over K and hence also over L. The theorem follows.

Theorem 27. Let V be an irreducible affine set defined over K, let $x \in V$, and let L be an extension field of K. Then x is simple on V if and only if it is simple on $V^{(L)}$.

Proof. By Lemma 13, the K-space $\mathrm{Der}_K(V, x)$ and the L-space $\mathrm{Der}_L(V^{(L)}, x)$ have the same dimension. The theorem follows because $V^{(L)}$ is irreducible and $\mathrm{Dim}\, V = \mathrm{Dim}\, V^{(L)}$.

Theorem 28. Let V be an irreducible affine set defined over K, and let L be an extension field of K. Suppose that $\eta \in V^{(L)}$ and is a generic point of V. Then η is simple on $V^{(L)}$.

Proof. We have an isomorphism $K[V] \xrightarrow{\sim} K[\eta]$ of K-algebras in which ξ_i is mapped into $\xi_i(\eta) = \eta_i$ (say). (This means that if $\hat{\xi}_i$ is the natural prolongation of ξ_i to $V^{(L)}$, then $\hat{\xi}_i(\eta) = \eta_i$.) Accordingly we have an isomorphism

$$K[\xi_1, \xi_2, \ldots, \xi_n] \approx K[\eta_1, \eta_2, \ldots, \eta_n]$$

of K-algebras which matches ξ_i with η_i. Hence, by Theorem 21 Cor., the matrix $\|\partial F_j/\partial\eta_i\|$ has rank $n - r$, where $r = \mathrm{Dim}\, V = \mathrm{Dim}\, V^{(L)}$. Consequently η is simple on $V^{(L)}$.

---

† We use the same notation as in (4.4.10).

4.5 Tangent spaces and products

This topic is treated in a separate section because it does not requi the auxiliary notation introduced at the beginning of section (4.4).

Let V, W be affine sets (defined over K) and suppose that $x \in V$, $y \in W$. Suppose that $D_1 \in \mathrm{Der}_K(V, x)$ and $D_2 \in \mathrm{Der}_K(W, y)$. If we mak the identification

$$K[V \times W] = K[V] \otimes_K K[W], \tag{4.5.1}$$

then there is a K-linear mapping

$$D : K[V \times W] \to K \tag{4.5.2}$$

in which

$$D(f \otimes g) = (D_1 f)g(y) + f(x)(D_2 g). \tag{4.5.3}$$

A simple verification shows that D belongs to $\mathrm{Der}_K(V \times W, (x, y))$. Thus we have a mapping

$$\mathrm{Der}_K(V, x) \oplus \mathrm{Der}_K(W, y) \to \mathrm{Der}_K(V \times W, (x, y)) \tag{4.5.4}$$

given by $(D_1, D_2) \mapsto D$.

Theorem 29. _The mapping (4.5.4) is an isomorphism of the K-space_ $\mathrm{Der}_K(V, x) \oplus \mathrm{Der}_K(W, y)$ _on to the K-space_ $\mathrm{Der}_K(V \times W, (x, y))$.

Proof. It is easily checked that (4.5.4) is K-linear and an injectio Now suppose that D is in $\mathrm{Der}_K(V \times W, (x, y))$. Define

$$D_1 : K[V] \to K \text{ and } D_2 : K[W] \to K$$

by $D_1 f = D(f \otimes 1)$ and $D_2 g = D(1 \otimes g)$. Then $D_1 \in \mathrm{Der}_K(V, x)$, $D_2 \in \mathrm{Der}_K(W, y)$ and D is the image, under (4.5.4), of (D_1, D_2).

Theorem 30. _Let_ V _and_ W _be irreducible affine sets and suppos that_ $x \in V$ _and_ $y \in W$. _Then_ (x, y) _is a simple point of_ $V \times W$ _if and only if_ x _is a simple point of_ V _and also_ y _is a simple point of_ W.

Proof. We know, from Chapter 3 Theorems 16 and 25, that $V \times W$ is irreducible and that its dimension is $\text{Dim } V + \text{Dim } W$. Next, by Theorem 29, the dimension of the K-space $\text{Der}_K(V \times W, (x, y))$ is the sum of the dimensions of $\text{Der}_K(V, x)$ and $\text{Der}_K(W, y)$. Theorem 30 now follows from Theorem 22 and the definition of a simple point.

4.6 Differentials

In this section we shall give another application of the theory of derivations. This will be useful when we come to study the Lie algebra of an affine group.

Throughout section (4.6) it will be assumed that K is _infinite._ In what follows A denotes an n-dimensional $(n \geq 1)$ vector space over K and we put $A^* = \text{Hom}_K(A, K)$ so that A^* consists of all the linear forms on A.

Lemma 14. _Let_ $\phi : A^* \to K[A]$ _be a_ K-_linear mapping. Then there exists a unique_ $D \in \text{Der}_K(K[A])$ _such that_ $DF = \phi(F)$ _for all_ $F \in A^*$.

Proof. Let $F_1, F_2, \ldots, F_n$ be a base for A^* over K. Then $K[A] = K[F_1, F_2, \ldots, F_n]$ and the F_i are algebraically independent over K. Now

$$D = \sum_{i=1}^{n} \phi(F_i) \frac{\partial}{\partial F_i}$$

belongs to $\text{Der}_K(K[A])$ and has the required property. Uniqueness is obvious.

Let $a \in A$ and take for $\phi : A^* \to K[A]$ the K-linear mapping in which $\phi(F) = F(a)$. By Lemma 14, this mapping gives rise to a derivation, Δ_a say, of $K[A]$ over K. Thus $\Delta_a \in \text{Der}_K(K[A])$ and

$$\Delta_a F = F(a) \tag{4.6.1}$$

for all $F \in A^*$. Naturally Δ_a extends to a derivation of $K(A)$ over K. This extension will be denoted by the same symbol.

Suppose that $H \in K(A)$ so that H is a rational function on A. If $b \in A$ and H is defined at b, then $\Delta_a H$ is also defined at b. Put

$$\mathrm{Def}(H) = \{b \mid b \in A \text{ and } H \text{ is defined at } b\} \tag{4.6.2}$$

and define

$$dH : \mathrm{Def}(H) \times A \to K \tag{4.6.3}$$

by

$$(dH)(b, a) = (\Delta_a H)(b). \tag{4.6.4}$$

The mapping dH is called the differential of H.

We record some properties of the differential. In the following formulae H, H_1, H_2 are rational functions on A, all defined at b, and $k \in K$. The formulae are:

$$(dH)(b, a_1 + a_2) = (dH)(b, a_1) + (dH)(b, a_2), \tag{4.6.5}$$

$$(dH)(b, ka) = k((dH)(b, a)), \tag{4.6.6}$$

$$(dkH)(b, a) = k((dH)(b, a)), \tag{4.6.7}$$

$$(d(H_1 + H_2))(b, a) = (dH_1)(b, a) + (dH_2)(b, a), \tag{4.6.8}$$

$$(d(H_1H_2))(b, a) = H_1(b)((dH_2)(b, a)) + H_2(b)((dH_1)(b, a)), \tag{4.6.9}$$

$$(dF)(b, a) = F(a) \text{ for all } F \in A^*. \tag{4.6.10}$$

Finally if $H = P/Q$, where $P, Q \in K[A]$ and $Q(b) \neq 0$, then

$$(dH)(b, a) = \frac{Q(b)((dP)(b, a)) - P(b)((dQ)(b, a))}{(Q(b))^2}. \tag{4.6.11}$$

We continue with the same assumptions but now we suppose that L is an extension field of K. Then, with the same notation as in section (1.6), A^L is an n-dimensional vector space over L. By Chapter 2 Theorem 32, $A^L = A^{(L)}$ because K is infinite.

Let $H \in K(A)$. Then H is a rational function on A and therefore it has a natural prolongation to a rational function, $\hat{H}$ say, on A^L. Note that if $F \in A^*$, so that F is a linear form on A, then its prolongation $\hat{F}$ is a linear form on A^L.

Suppose now that $a \in A \subseteq A^L$. Then we can form $\Delta_a \hat{F}$ and $\Delta_a F$. It is clear that $\Delta_a \hat{F}$ is the prolongation of $\Delta_a F$. We conclude at once from this that $\Delta_a \hat{H}$ is the prolongation of $\Delta_a H$ and therefore

$$(d\hat{H})(b, a) = (dH)(b, a) \tag{4.6.12}$$

provided that $b \in A$ and H is defined at b.

Now assume that $D \in \mathrm{Der}_K(L)$. Choose a base $a_1, a_2, \ldots, a_n$ for A over K. Then these same elements will constitute a base for A^L over L. Hence if $x \in A^L$ we can write x in the form $x = \lambda_1 a_1 + \lambda_2 a_2 + \ldots + \lambda_n a_n$ with $\lambda_i \in L$. Put

$$Dx = (D\lambda_1)a_1 + (D\lambda_2)a_2 + \ldots + (D\lambda_n)a_n. \tag{4.6.13}$$

Then when x, x_1, x_2 belong to A^L and $\lambda \in L$, we have

$$D(x_1 + x_2) = Dx_1 + Dx_2, \tag{4.6.14}$$

$$D(\lambda x) = (D\lambda)x + \lambda(Dx). \tag{4.6.15}$$

Also

$$Db = 0 \text{ for all } b \in A. \tag{4.6.16}$$

In particular it follows that, when $\lambda_1, \lambda_2, \ldots, \lambda_m$ belong to L and $b_1, b_2, \ldots, b_m$ to A,

$$D(\lambda_1 b_1 + \lambda_2 b_2 + \ldots + \lambda_m b_m) = \sum_{i=1}^{m} (D\lambda_i)b_i. \tag{4.6.17}$$

This shows that *the definition of* Dx *as provided by (4.6.13) is independent of the choice of the* K-*base* $a_1, a_2, \ldots, a_n$ *of* A. Note too that

$$\hat{F}(Dx) = D(\hat{F}(x)), \tag{4.6.18}$$

for all $F \in A^*$.

Lemma 15. *Let* $H \in K(A)$ *and let* $\hat{H}$ *be its prolongation to a rational function on* A^L. *Let* $D \in \mathrm{Der}_K(L)$. *Suppose that* $x \in A^L$ *and that* $\hat{H}$ *is defined at* x. *Then*

$$D(\hat{H}(x)) = (d\hat{H})(x, Dx).$$

Proof. Let R_x consist of all the rational functions on A whose prolongations are defined at x. Evidently R_x is a subalgebra of the K-algebra K(A). Let the mapping $\phi : R_x \to L$ be given by $\phi(S) = \hat{S}(x)$. This is a homomorphism of K-algebras.

Define $\Omega : R_x \to L$ and $\overline{\Omega} : R_x \to L$ by

$$\Omega(S) = (d\hat{S})(x, Dx)$$

and

$$\overline{\Omega}(S) = D(\hat{S}(x)).$$

We wish to show that Ω and $\overline{\Omega}$ coincide. Clearly they are additive and they both vanish on K. Also if $S_1, S_2 \in R_x$, then

$$\Omega(S_1 S_2) = \phi(S_1)\Omega(S_2) + \phi(S_2)\Omega(S_1)$$

and

$$\overline{\Omega}(S_1 S_2) = \phi(S_1)\overline{\Omega}(S_2) + \phi(S_2)\overline{\Omega}(S_1).$$

Again if $F \in A^*$, then $F \in R_x$ and, by (4.6.10),

$$\Omega(F) = \hat{F}(Dx)$$

which is equal to $D(\hat{F}(x)) = \overline{\Omega}(F)$ by (4.6.18). It follows that Ω and $\overline{\Omega}$ agree on K[A].

Suppose that $S \in R_x$. We can choose[†] P, Q $\in$ K[A] so that $S = P/$ and $\hat{Q}(x) \neq 0$. By (4.6.11), $\Omega(S)$ is equal to

$$\begin{aligned} &\frac{\hat{Q}(x)((d\hat{P})(x, Dx)) - \hat{P}(x)((d\hat{Q})(x, Dx))}{(\hat{Q}(x))^2} \\ &= \frac{\hat{Q}(x)(D\hat{P}(x)) - \hat{P}(x)(D\hat{Q}(x))}{(\hat{Q}(x))^2} \\ &= D(\hat{P}(x)/\hat{Q}(x)) \\ &= D(\hat{S}(x)) \\ &= \overline{\Omega}(S). \end{aligned}$$

† C.f. the proof of Theorem 31 of Chapter 3.

The lemma follows.

We continue to assume that K is an infinite field but now suppose that A is a non-trivial, unitary, associative K-algebra whose dimension as a K-space is finite. Let $a \in A$ and define $\lambda_a : A \to A$ by $\lambda_a(b) = ab$. This mapping induces a linear mapping $\lambda_a^* : A^* \to A^*$ of K-spaces and now we can define a K-linear mapping $\phi : A^* \to K[A]$ by $\phi(F) = -\lambda_a^*(F)$. Lemma 14 shows that ϕ gives rise to a derivation, D_a say, of $K[A]$ over K. Thus $D_a \in \mathrm{Der}_K(K[A])$ and

$$D_a F = -(F \circ \lambda_a) \tag{4.6.19}$$

for all $F \in A^*$. Accordingly

$$(D_a F)(b) = -F(ab) \tag{4.6.20}$$

provided that $F \in A^*$ and $a, b \in A$.

Next we establish a connection between the derivation D_a and the notion of a differential. Observe first that D_a extends to a derivation of $K(A)$ over K (we use D_a also to denote the extension) and if $H \in K(A)$ is defined at b, where $b \in A$, then $D_a H$ is defined at b as well.

Theorem 31. Let H be a rational function on A which is defined at b. Then

$$(D_a H)(b) = -(dH)(b, ab)$$

for all $a \in A$.

Proof. Denote by $\mathfrak{A}$ the set of all P, in K[A], for which

$$(D_a P)(b) = -(dP)(b, ab).$$

(Here a and b are kept fixed.) It is clear that $K \subseteq \mathfrak{A}$ and, by (4.6.20) and (4.6.10), we see that $A^* \subseteq \mathfrak{A}$ as well.

Next assume that $P_1, P_2 \in \mathfrak{A}$. An easy verification shows that $P_1 + P_2$ and $P_1 P_2$ also belong to $\mathfrak{A}$. This shows that $\mathfrak{A} = K[A]$.

Finally we can write H in the form P/Q, where $P, Q \in K[A]$ and $Q(b) \neq 0$. If we now use (4.6.11) and the fact that D_a is a derivation we obtain the desired result.

We add a few remarks about prolongations. Let L be an extension field of K, then A^L is a unitary, associative, L-algebra.† As before if $H \in K(A)$, then we use $\hat{H}$ to denote its prolongation to a rational function on A^L. If now $F \in A^*$ and $a, b \in A$, then

$$(D_a \hat{F})(b) = -\hat{F}(ab) = -F(ab) = (D_a F)(b).$$

Thus $D_a \hat{F}$ is the prolongation of $D_a F$. It follows immediately that _for any_ $H \in K(A)$, $D_a \hat{H}$ _is the prolongation of_ $D_a H$.

The results of section (4.6) will find applications in Chapter 6.

† See section (1.6).

Part II. Affine Groups

5 · Affine groups

General remarks

Throughout Chapter 5, K will denote an arbitrary field and if X is a subset of a topological space, then $\overline{X}$ will be used to denote its closure.

In the case of a group G, the letter e is used to denote its identity element. Should it be necessary to be more explicit the symbol e_G will be employed. In the case of an affine group (see section (5.1)) we shall always use G_0 to denote the connected component containing the identity element.

5.1 Affine groups

Let G be an affine set and suppose that G also has the structure of an abstract (multiplicative) group. Consider the mappings

$$G \times G \to G \text{ and } G \to G$$

in which $(x, y) \mapsto xy$ and $x \mapsto x^{-1}$ respectively. If both of these are K-morphisms, then it would be perfectly reasonable to describe the whole situation by saying that G is an affine group. However it will be convenient to give an equivalent definition in rather different terms. We therefore make a fresh start.

Let G be an affine set (defined over K) and suppose that we are given K-morphisms

$$\mu : G \times G \to G, \tag{5.1.1}$$

$$j : G \to G, \tag{5.1.2}$$

and a point e of G,

Definition. The quadruplet $[G, \mu, j, e]$ is called an 'affine group' provided that

(i) $\mu(\mu(x, y), z) = \mu(x, \mu(y, z))$,

(ii) $\mu(x, e) = x = \mu(e, x)$,

(iii) $\mu(x, j(x)) = e = \mu(j(x), x)$,

for all x, y, z in G.

Suppose that we have this situation. Then G is a group with μ as its law of composition, j the operation of taking inverses, and e the identity element. We normally, and without special comment, use the multiplicative notation, that is we write $\mu(x, y) = xy$ and $j(x) = x^{-1}$. However we shall translate certain results so as to make addition the law of composition on the few occasions when this is more convenient.

Let $[G, \mu, j, e]$ be an affine group. We can construct two K-morphisms

$$G \times G \times G \to G$$

as follows: in the first $(x, y, z) \mapsto \mu(\mu(x, y), z)$ and in the second $(x, y, z) \mapsto \mu(x, \mu(y, z))$. Condition (i) says that these coincide. Similarl (ii) says that the K-morphisms $x \mapsto \mu(x, e)$ and $x \mapsto \mu(e, x)$ are the sam as the identity morphism $G \to G$. Finally (iii) asserts that the K-morphis $x \to \mu(x, j(x))$ and $x \to \mu(j(x), x)$ both agree with the constant K-morphis $G \to G$ in which every point is mapped into e.

One advantage of presenting the definition in this way is that it enables us to show that the property of being of an affine group is preserved under an extension of the ground field.

To see how this arises let us assume that $[G, \mu, j, e]$ is an affine group and that L is an extension field of K. By (2.10.9), $(G \times G)^{(L)} = G^{(L)} \times G^{(L)}$ and therefore $\mu^{(L)}$ is an L-morphism

$$\mu^{(L)} : G^{(L)} \times G^{(L)} \to G^{(L)}.$$

Theorem 1. _Let_ $[G, \mu, j, e]$ _be an affine group defined over_ K, _and let_ L _be an extension field of_ K. _Then_ $[G^{(L)}, \mu^{(L)}, j^{(L)}, e]$ _is an affine group defined over_ L.

Proof. Define L-morphisms

$$p : G^{(L)} \times G^{(L)} \times G^{(L)} \to G^{(L)}$$

and

$$q : G^{(L)} \times G^{(L)} \times G^{(L)} \to G^{(L)}$$

by $p(\xi, \eta, \zeta) = \mu^{(L)}(\mu^{(L)}(\xi, \eta), \zeta)$ and $q(\xi, \eta, \zeta) = \mu^{(L)}(\xi, \mu^{(L)}(\eta, \zeta))$. Since $G^{(L)} \times G^{(L)} \times G^{(L)} = (G \times G \times G)^{(L)}$ and since p and q extend the same K-morphism

$$G \times G \times G \to G,$$

it follows, by Chapter 2 Lemma 15 Cor., that $p = q$. This verifies the first of the three conditions in the definition of an affine group and the other two may be checked similarly.

Corollary 1. Let G be an affine group defined over K, and let L be an extension field of K. Then there is just one way to endow the affine set $G^{(L)}$ with the structure of an affine group so that G becomes a subgroup.

Proof. We use the same notation. Let $[G^{(L)}, \hat{\mu}, \hat{j}, e]$ be an affine group having G as a subgroup. It will suffice to show that $\hat{\mu} = \mu^{(L)}$. Now the L-morphism

$$\hat{\mu} : (G \times G)^{(L)} \to G^{(L)}$$

extends the K-morphism $\mu : G \times G \to G$ by hypothesis and therefore $\hat{\mu} = \mu^{(L)}$ as required.

Corollary 2. If G is a commutative affine group, then so is $G^{(L)}$.

Proof. Define an L-morphism $\phi : G^{(L)} \times G^{(L)} \to G^{(L)}$ by $\phi(\xi, \eta) = \mu^{(L)}(\eta, \xi)$. This extends the K-morphism $\mu : G \times G \to G$ so $\phi = \mu^{(L)}$. The corollary follows.

The notation $[G, \mu, j, e]$ is more or less indispensable for the proper statement of Theorem 1, but it is too cumbersome for general use.

Frequently we shall suppress any direct reference to μ, j, and e and say simply that G is an affine group. Also *for the rest of section (5.1) we shall use the multiplicative notation and employ* xy *as an alternative to* $\mu(x, y)$.

Let G be an affine group (defined over K) and let $x \in G$. The mapping

$$\lambda_x : G \to G \tag{5.1.3}$$

in which $\lambda_x(y) = xy$ is a K-morphism called *left translation* by means of x. In fact λ_x is a K-isomorphism (of the affine set G on to itself) whose inverse is $\lambda_{x^{-1}}$. The *right translation* by means of x is the K-morphism

$$\rho_x : G \to G \tag{5.1.4}$$

given by $\rho_x(y) = yx$. This too is a K-isomorphism of the affine set G on to itself.

The next theorem asserts that an affine group, when regarded as an affine set, is homogeneous.

Theorem 2. *Let* G *be an affine group (defined over* K) *and let* x, y *belong to* G. *Then there exists a* K-*isomorphism* ϕ, *of the affine set* G *on to itself, such that* $\phi(x) = y$.

In fact we may take ϕ to be the left translation by yx^{-1}.

We next turn our attention to the closed subgroups of an affine group. In preparation for this we prove a general lemma.

Lemma 1. *Let* V, W *be affine sets defined over* K *and let* $\phi : V \to W$ *and* $\psi : V \to W$ *be* K-*morphisms. Then*

$$\{x \mid x \in V \text{ and } \phi(x) = \psi(x)\}$$

is a closed subset of V.

Proof. Define a K-morphism $p : V \to W \times W$ by $p(x) = (\phi(x), \psi(x))$. Then

$$\{x \mid x \in V \text{ and } \phi(x) = \psi(x)\} = p^{-1}(\Delta),$$

where Δ denotes the diagonal of $W \times W$. The lemma follows because, by section (2.9), Δ is closed in $W \times W$.

Now let $[G, \mu, j, e]$ be an affine group and let H be a closed subgroup of G, that is H is a closed subset of the affine set G and a subgroup of the abstract group G. By restriction μ and j induce K-morphisms

$$\mu' : H \times H \to H$$

and

$$j' : H \to H$$

and it is clear that $[H, \mu', j', e]$ is an affine group. We embody these observations in

Theorem 3. Let G be an affine group defined over K and let H be a closed subgroup of G. Then H has an induced structure as an affine group defined over the same field.

From here on, every closed subgroup of an affine group will automatically be regarded as an affine group. Suppose that H is a closed subgroup of the affine group G and let L be an extension field of K. Then on the one hand $H^{(L)}$ is an affine group with $\mu'^{(L)}$ as its law of composition (we are retaining our previous notation), and on the other it is a closed subset of $G^{(L)}$. Using the inclusion mapping $H^{(L)} \to G^{(L)}$ we can construct, in an obvious manner, the L-morphisms

$$H^{(L)} \times H^{(L)} \xrightarrow{\mu'^{(L)}} H^{(L)} \longrightarrow G^{(L)}$$

and

$$H^{(L)} \times H^{(L)} \to G^{(L)} \times G^{(L)} \xrightarrow{\mu^{(L)}} G^{(L)}.$$

Since these extend the same K-morphism $H \times H \to G$, it follows, again by Chapter 2 Lemma 15 Cor., that they coincide. Accordingly the inclusion mapping $H^{(L)} \to G^{(L)}$ is a homomorphism of groups.

Corollary. <u>Let</u> G <u>be an affine group defined over</u> K <u>and let</u> H <u>be a closed, normal subgroup of</u> G. <u>If now</u> L <u>is an extension field of</u> K, <u>then</u> $H^{(L)}$ <u>is a closed normal subgroup of</u> $G^{(L)}$.

Proof. Let $h \in H$. Then

$$\{\xi \mid \xi \in G^{(L)} \text{ and } \xi h \xi^{-1} \in H^{(L)}\}$$

is the inverse image of $H^{(L)}$ with respect to the L-morphism $\xi \mapsto \xi h \xi^{-1}$. The inverse image is therefore closed in $G^{(L)}$ and so, since it contains G, it is $G^{(L)}$ itself. Thus $\xi h \xi^{-1} \in H^{(L)}$ for $\xi \in G^{(L)}$ and $h \in H$.

Now let $\xi \in G^{(L)}$. This time

$$\{\eta \mid \eta \in G^{(L)} \text{ and } \xi \eta \xi^{-1} \in H^{(L)}\}$$

is closed in $G^{(L)}$ and it contains H. Consequently, by Chapter 2 Theorem 30, it must contain $H^{(L)}$. Accordingly $\xi \eta \xi^{-1} \in H^{(L)}$ for all $\xi \in G^{(L)}$ and $\eta \in H^{(L)}$, so the corollary is proved.

Let us investigate further the question of closed subgroups. Suppose that G is an affine group and assume that $\sigma \in G$. Put

$$C_\sigma = \{x \mid x \in G \text{ and } \sigma x = x\sigma\}. \tag{5.1.5}$$

Lemma 2. C_σ <u>is a closed subgroup of</u> G.

Proof. Let λ_σ respectively ρ_σ denote the left respectively right translation by means of σ. Then

$$C_\sigma = \{x \mid x \in G \text{ and } \lambda_\sigma(x) = \rho_\sigma(x)\}.$$

Consequently, by Lemma 1, C_σ is a closed subset of G. It is obvious that it is a subgroup of G.

Theorem 4. <u>Let</u> G <u>be an affine group. Then its centre is a closed subgroup of</u> G.

Proof. Define C_σ as in (5.1.5). The intersection of all the C_σ's is the centre of G. The theorem therefore follows from Lemma 2.

Theorem 5. *Let* G *be an affine group and* H *a subgroup of* G. *If* $\overline{H}$ *is the closure of* H *in* G, *then* $\overline{H}$ *is also a subgroup of* G.

Proof. The mapping $x \mapsto x^{-1}$ is a *homeomorphism* of the affine set G on to itself. Consequently $(\overline{H})^{-1}$ is the closure of $H^{-1} = H$ and therefore $(\overline{H})^{-1} = \overline{H}$. Thus $\overline{H}$ is closed under the taking of inverses.

Let $h \in H$ and consider left translation by means of h. Since this also provides a homeomorphism, $h\overline{H}$ is the closure of $hH = H$, that is $h\overline{H} = \overline{H}$.

Finally assume that $x \in \overline{H}$. Then $\overline{H}x$ is the closure of Hx and $Hx \subseteq H\overline{H} = \overline{H}$. Accordingly $\overline{H}x \subseteq \overline{H}$ which shows that $\overline{H}$ is also closed under multiplication. The theorem follows.

If we are given a subset of an affine group G, then there will be a smallest closed subgroup of G which contains the given subset. What follows now is preparation for an investigation of the connection between the subset and the closed subgroup which it generates.

Let $x \in G$. We have already remarked that the right translation $\rho_x : G \to G$ is a K-isomorphism of affine sets. It therefore induces an isomorphism

$$\rho_x^* : K[G] \to K[G] \tag{5.1.6}$$

of K-algebras. Note that ρ_e^* is the identity mapping of $K[G]$ and that

$$\rho_x^* \circ \rho_y^* = \rho_{xy}^* \tag{5.1.7}$$

for all $x, y \in G$. Also if $f \in K[G]$, then

$$(\rho_x^*(f))(y) = f(yx). \tag{5.1.8}$$

Theorem 6. *Let* G *be an affine group defined over* K, *and let* S *be a non-empty multiplicatively closed subset of* G. *Further let* $f \in K[G]$ *and denote by* U_f *the K-subspace of* $K[G]$ *generated by the family* $\{\rho_z^*(f)\}_{z \in S}$. *Then the dimension of* U_f *is finite and* $\rho_z^*(U_f) = U_f$ *for all* $z \in S$.

Remark. There is, of course, a similar result based on left translations, but we shall not state it separately.

Proof. Define $\omega : G \times G \to K$ by

$$\omega(y, x) = f(yx) = (\rho_x^*(f))(y).$$

Then $\omega \in K[G \times G]$ and therefore there exist $f_1, f_2, \ldots, f_m$ and $g_1, g_2, \ldots, g_m$ in $K[G]$ such that

$$\omega(y, x) = \sum_{i=1}^{m} f_i(y)g_i(x)$$

for all x, y in G. Thus

$$\rho_z^*(f) = \sum_{i=1}^{m} g_i(z)f_i$$

which proves the first assertion. Next when $z, x \in S$ we have

$$\rho_z^*(\rho_x^*(f)) = \rho_{zx}^*(f) \in U_f$$

which shows that $\rho_z^*(U_f) \subseteq U_f$. Consequently, when $z \in S$, ρ_z^* induces an endomorphism of the K-space U_f. Since the dimension of U_f is finite and this particular endomorphism is an injection, we see that $\rho_z^*(U_f) = U_f$ as required.

Theorem 7. _Let_ G _be an affine group (defined over_ K) _and let_ U _be a subspace of_ $K[G]$ _when_ $K[G]$ _is considered as a vector space over_ K. _Suppose that_ $x \in G$ _is such that_ $\rho_x^*(U) \subseteq U$. _Then_ $\rho_x^*(U) = U$.

Remark. Naturally there is an analogous result for left translations.

Proof. Let S consist of e and the elements x^n, where $n \geq 1$. Then S is a multiplicatively closed subset of G and $\rho_z^*(U) \subseteq U$ for all $z \in S$. Let $f \in U$ and define U_f as in Theorem 6. Then

$$f \in U_f = \rho_x^*(U_f) \subseteq \rho_x^*(U).$$

The theorem follows.

The next result is very striking.[†]

Theorem 8. Let G be an affine group and S a non-empty multiplicatively closed subset of G. Then the closure $\bar{S}$, of S in G, is the smallest closed subgroup of G containing S.

Proof. It is enough to show that $\bar{S}$ is a subgroup of G. Put $\mathfrak{U} = I_G(S) = I_G(\bar{S})$ and let $f \in \mathfrak{U}$. The mapping $\omega : G \times G \to K$ given by

$$\omega(y, x) = f(yx) = (\rho_x^*(f))(y)$$

belongs to $K[G \times G]$ and it vanishes on $S \times S$ because S is multiplicatively closed. It must therefore vanish on the closure of $S \times S$ in $G \times G$. However, by Chapter 2 Theorem 28 Cor., this closure is $\bar{S} \times \bar{S}$. Hence if $\tau, \sigma \in \bar{S}$, then

$$(\rho_\sigma^*(f))(\tau) = \omega(\tau, \sigma) = 0$$

whence $\rho_\sigma^*(f) \in \mathfrak{U}$. This shows that $\rho_\sigma^*(\mathfrak{U}) \subseteq \mathfrak{U}$. It follows, from Theorem 7, that $\rho_\sigma^*(\mathfrak{U}) = \mathfrak{U}$. In particular $f = \rho_\sigma^*(f')$ for some $f' \in \mathfrak{U}$. Accordingly $f(e) = f'(\sigma) = 0$. Thus to sum up: $\rho_\sigma^*(\mathfrak{U}) = \mathfrak{U}$ for all $\sigma \in \bar{S}$ and furthermore $e \in \bar{S}$.

Next suppose that $x \in G$ and $\rho_x^*(\mathfrak{U}) = \mathfrak{U}$. Then $\mathfrak{U} = \rho_{x^{-1}}^*(\mathfrak{U})$ and so for $f \in \mathfrak{U}$ we can choose $f_1 \in \mathfrak{U}$ so that $f = \rho_{x^{-1}}^*(f_1)$. It follows that $f(x) = f_1(e) = 0$ and therefore

$$x \in C_G(\mathfrak{U}) = \bar{S}.$$

Combining our results so far we find that

$$\bar{S} = \{x \mid x \in G \text{ and } \rho_x^*(\mathfrak{U}) = \mathfrak{U}\}$$

and now it is a trivial matter to check that $\bar{S}$ is a subgroup of G.

---

† See C. Chevalley [(2) Proposition 2, p. 82].

Corollary. Let G and S be as in the theorem and put $\mathfrak{U}=I_G(S)$. Then

$$\{x \mid x \in G \text{ and } \rho^*_x(\mathfrak{U}) = \mathfrak{U}\}$$

is the smallest closed subgroup of G containing S.

The above proof of Theorem 8 is somewhat indirect but it has the advantage of introducing ideas that will be useful in other contexts. A short self-contained proof of the theorem will now be given.†

Assume that the hypotheses of Theorem 8 are satisfied. By adapting the proof of Theorem 5 we can readily check that $\overline{S}$ is multiplicatively closed. We may therefore suppose that S itself is closed and seek to prove that S is a group.

Let $x \in S$. The collection of subsets of S that are (i) topologically and multiplicatively closed, and (ii) contain a positive power of x is non-empty. In fact it has a minimal member M (say) because G is a Noetherian topological space. By construction $x^n \in M$ for some $n \geq 1$.

Consider $x^n M$. This is topologically closed, because left translations are homeomorphisms, and it is clearly multiplicatively closed as well. Also $x^n M \subseteq M$, because M is multiplicatively closed, and $x^n M$ contains x^{2n}. Consequently, by the choice of M, $x^n M = M$ and therefore $M = x^{2n} M$. This shows that $x^n = x^{2n} y$ for some $y \in M \subseteq S$ and now we have $x^n y = e$. It follows that $x^{-1} \in S$ and with this the theorem is proved.

We return to the general discussion in order to record some simple observations on direct products of affine groups. To this end let G_1, G_2 be affine groups defined over K. Then $G_1 \times G_2$ is an affine set and it has a group structure. Indeed it is readily verified that $G_1 \times G_2$ is an affine group. More generally, if $G_1, G_2, \ldots, G_m$ are affine groups, then the direct product

$$G_1 \times G_2 \times \ldots \times G_m$$

is also an affine group. Next let L be an extension field of K. Then

† The argument which follows was communicated to me by **P. Vámos.**

Theorem 1 Cor. 1 shows that we can identify

$$(G_1 \times G_2 \times \ldots \times G_m)^{(L)}$$

and

$$G_1^{(L)} \times G_2^{(L)} \times \ldots \times G_m^{(L)}$$

as affine groups in the obvious way.

5.2 Components of an affine group

Throughout section (5.2), G will denote an affine group defined over K, and we shall use the multiplicative notation. Since G is a space by virtue of its affine topology, it is the disjoint union of its connected components. We shall normally use G_0 to denote the particular connected component that contains the identity element e of G. It is customary to refer to G_0 as the connected component of the identity though this is an abuse of language.

Theorem 9. The connected component G_0 that contains the identity element of G is a closed normal subgroup of G.

Proof. The homeomorphism $x \mapsto x^{-1}$ shows that G_0^{-1} is a connected component of G containing e. Hence $G_0^{-1} = G_0$. Let $z \in G_0$. Then $z^{-1} \in G_0$ and now the left translation of G by means of z shows that zG_0 is a connected component of G containing $zz^{-1} = e$. Consequently $zG_0 = G_0$ and we have proved that G_0 is a subgroup of G. It must be a closed subgroup because the connected components of a topological space are always closed.

Let $x \in G$ and define $p : G \to G$ by $p(y) = xyx^{-1}$. Then p is compounded from two translations and is therefore a homeomorphism. This makes it clear that $p(G_0) = G_0$, that is to say G_0 is a normal subgroup of G.

We shall next examine the connection between G_0 and the irreducible components, $V_1, V_2, \ldots, V_r$ say, of G. We arrange the numbering so that $e \in V_1$. Note that $\{V_1^{-1}, V_2^{-1}, \ldots, V_r^{-1}\}$ is a permutation of

$\{V_1, V_2, \ldots, V_r\}$ and so too are

$$\{zV_1, zV_2, \ldots, zV_r\} \text{ and } \{V_1z, V_2z, \ldots, V_rz\}$$

for every $z \in G$.

Since G is the irredundant union of $V_1, V_2, \ldots, V_r$, we can choose $x \in V_1$ so that $x \notin V_2, \ldots, x \notin V_r$. This secures that, for each $y \in G$, xy belongs to just one of $V_1, V_2, \ldots, V_r$.

Suppose that $1 \leq i \leq r$. Then $x^{-1}V_i = V_k$ for some k. Let us determine k. Since $V_1y = V_i$ if and only if $xy \in V_i$, that is to say if and only if $y \in V_k$, it follows that $V_1V_k \subseteq V_i$. However $e \in V_1$ and therefore we may conclude that $k = i$. Accordingly $V_1V_i = V_i$ for $i = 1, 2, \ldots, r$. In this relation replace V_i by V_1^{-1}. This shows that $V_1V_1^{-1} = V_1^{-1}$ and hence that $V_1V_1^{-1} = V_1$. It has now been proved that V_1 is a subgroup of G and, because it is connected, we have $V_1 \subseteq G_0$. We recall that each of $V_1, V_2, \ldots, V_r$ is closed.

Assume that $2 \leq i \leq r$ and choose $y_i \in V_i$ so that it does not belong to any other irreducible component. Then $V_1y_i = V_i$ and thus we see that the irreducible components of G are none other than the right cosets of V_1. In particular, they are disjoint. However G_0 is connecte and it contains e. Accordingly G_0 does not meet $V_2 \cup V_3 \cup \ldots \cup V_r$ and therefore we must have $G_0 = V_1$. This proves

Theorem 10. *Let G be an affine group and G_0 the connected component of the identity element. Then the irreducible components of G are just the cosets of G_0 in G. Thus the irreducible components of G are disjoint, and the index of G_0 in G is finite.*

Corollary. *Any two irreducible components of the affine group G are K-isomorphic and so have the same dimension. This is equal to the dimension of G. In particular $\operatorname{Dim} G_0 = \operatorname{Dim} G$.*

Theorem 11. *Let H be a closed subgroup (of the affine group G) whose index in G is finite. Then $G_0 \subseteq H$.*

Proof. The (left) cosets of H in G are closed. Consequently H and $G \backslash H$ are disjoint closed subsets of G. Since $e \in H$ and G_0 is connected, we must have $G_0 \subseteq H$.

An important fact which is an immediate consequence of Theorem 10 is stated separately as

Theorem 12. Let G be an affine group. Then the following two statements are equivalent:

(a) as an affine set G is connected;

(b) as an affine set G is irreducible.

Thus for affine groups connected and irreducible amount to the same thing. In what follows we shall find it more convenient to speak of connected rather than irreducible groups.

Theorem 13. Let G be a connected affine group. Then every point of G is a simple point.

Remark. This result is sometimes described by saying that connected affine groups are smooth.

Proof. Since G is irreducible, Theorem 24 of Chapter 4 shows that it has at least one simple point. The fact that all of its points are simple follows by combining Theorem 2 with Theorem 25 of Chapter 4.

Theorem 14. Let G be a connected affine group and let U, V be non-empty open subsets of G. Then $G = UV$.

Proof. Let $x \in G$. Then U and xV^{-1} are non-empty open subsets of G and, furthermore, G is irreducible. Hence, by Chapter 3 Theorem 3, there exists $u \in U$ which also belongs to xV^{-1}. Accordingly $x = uv$ for some $v \in V$.

Corollary. If U is a non-empty open subset of a connected affine group G, then $G = UU$.

The next result has important applications in the theory of solvable affine groups.

Theorem 15. Let G be an affine group and let the ground field K be algebraically closed. Suppose that $S_1, S_2, \ldots, S_r$ are subsets of G having the following properties:

(i) $e \in S_i$ for $i = 1, 2, \ldots, r$;

(ii) for each i the closure $\bar{S}_i$, of S_i in G, is irreducible;

(iii) for each i, S_i contains a non-empty open subset of $\bar{S}_i$.

If now S is the smallest multiplicatively closed subset of G containing $S_1, S_2, \ldots, S_r$, then S is a closed connected subgroup of G.

Remark. Condition (iii) means that there is a non-empty subset of $\bar{S}_i$ which is open in the topology of $\bar{S}_i$ that is contained in S_i.

Proof. Put $W = \bar{S}_1 \times \bar{S}_2 \times \ldots \times \bar{S}_r$. Then W is an irreducible affine set and it contains

$$W_0 = S_1 \times S_2 \times \ldots \times S_r.$$

For each positive integer n put

$$W^n = W \times W \times \ldots \times W,$$

$$W_0^n = W_0 \times W_0 \times \ldots \times W_0,$$

where in each case the product contains n factors. Now although the affine topology on a product is different from the ordinary product topolog none the less W_0^n contains a non-empty open subset of W^n. But W^n is a product

$$\bar{S}_1 \times \ldots \times \bar{S}_r \times \bar{S}_1 \times \ldots \times \bar{S}_r \times \ldots \times \bar{S}_1 \times \ldots \times \bar{S}_r$$

so each element of W^n is a sequence of elements of G. Let $W^n \to G$ be the K-morphism obtained by multiplying together the terms of the sequence without disturbing their order. Then if $W^{[n]}$ denotes the image of W^n and $W_0^{[n]}$ that of W_0^n, we arrive at an almost surjective K-morphism $W^n \to W^{[n]}$ of affine sets. Note that W^n (and hence also $W^{[n]}$) is irreducible and therefore, by Chapter 3 Theorem 33, $W_0^{[n]}$ contains a non-empty open subset of $W^{[n]}$.

Since $e \in S_i$ for $i = 1, 2, \ldots, r$, we have $W^{[n]} \subseteq W^{[n+1]}$. Thus

$$\overline{W^{[1]}} \subseteq \overline{W^{[2]}} \subseteq \overline{W^{[3]}} \subseteq \ldots \subseteq G$$

and each term in the sequence is irreducible. It follows that there exists a positive integer t such that

$$\overline{W^{[n]}} = \overline{W^{[t]}}$$

for all $n \geq t$. Now S is contained in the union of the $W_0^{[n]}$ and therefore $S \subseteq \overline{W^{[t]}}$. Thus $S \subseteq \overline{S} \subseteq \overline{W^{[t]}}$. Also $W_0^{[t]}$ contains a non-empty open subset of $\overline{W^{[t]}}$ and therefore this is also true of S. Accordingly, by Chapter 3 Theorem 2, $\overline{S} = \overline{W^{[t]}}$. This shows that $\overline{S}$ is connected and, by Theorem 8, it is a closed subgroup of G. However we know that S contains a non-empty open subset of $\overline{W^{[t]}} = \overline{S}$. It therefore follows, from Theorem 14 Cor., that $\overline{S} = SS = S$. The proof is now complete.

Corollary 1. *Let the ground field* K *be algebraically closed, let* W *be an irreducible affine set, and let* $\phi : W \to G$ *be a* K*-morphism of affine sets, where* G *is an affine group. If now* $e \in \phi(W)$, *then the smallest multiplicatively closed subset of* G *containing* $\phi(W)$ *is a closed, connected subgroup of* G.

Proof. Put $S_1 = \phi(W)$. Then conditions (i) and (ii) of the theorem hold trivially, whereas condition (iii) holds by virtue of Chapter 3 Theorem 33. The corollary follows.

We recall that if H is any group, then the *commutator subgroup* (H, H) of H is the subgroup that one obtains by taking all finite products of elements of the form $\sigma\tau\sigma^{-1}\tau^{-1}$, where σ and τ belong to H. This is a normal subgroup of H. In fact it is the smallest normal subgroup of G with an abelian factor group.

Corollary 2. *Let the affine group* G *be connected and let* K *be algebraically closed. Then the commutator subgroup* (G, G) *is a closed connected subgroup of* G.

Proof. The product $G \times G$ is irreducible and the mapping $\phi : G \times G \to G$ in which $\phi(\sigma, \tau) = \sigma\tau\sigma^{-1}\tau^{-1}$ is a K-morphism. Also $e \in \phi(G \times G)$ and the smallest multiplicatively closed subset of G containing $\phi(G \times G)$ is (G, G). The desired result therefore follows from Corollary 1.

5.3 Examples of affine groups

All the affine groups considered in section (5.3) will be defined over K. If G is a group, then $\mu : G \times G \to G$ will denote the compositio mapping (multiplication or addition as the case may be) and $j : G \to G$ the mapping which takes each element into its inverse.

Example 1. Let G be a finite group. Then G is an affine set and Theorem 18 of Chapter 2 shows that $[G, \mu, j, e]$ is an affine group. G is connected if and only if e is the only element of G.

Example 2. Let V be an n-dimensional vector space over K, put† $E = \mathrm{End}_K(V)$, and for each $g \in E$ denote by $D(g)$ the determinant of g. If now

$$GL(V) = \{g \mid g \in E \text{ and } D(g) \neq 0\},$$

then GL(V) is a group and, as was shown in section (2.5), it has a natural structure as an affine set.

Suppose first that K is a finite field. Then GL(V) is a finite grou and therefore it is an affine group by Example 1.

Now assume that K is infinite. The remarks made at the end of section (2.5) show that

$$\mu : GL(V) \times GL(V) \to GL(V)$$

and

$$j : GL(V) \to GL(V)$$

are K-morphisms. Hence GL(V) is again an affine group.

It is usual to call GL(V) the general linear group of V. This has a particularly important role in the theory of affine groups.‡

We add a little extra information in

---

† See section (2.5) Example 5.

‡ See Theorem 27.

Theorem 16. _Let_ V _be an_ n-_dimensional vector space over an infinite field_ K. _Then_ $GL(V)$ _is a connected affine group whose dimension is_ n^2.

That $GL(V)$ is connected and has dimension n^2 are consequences of (2.5.7). In fact the formula quoted shows that $K(GL(V))$ is a pure transcendental extension of K.

Example 3. Let $n \geq 1$ be an integer and let $GL_n(K)$ be the multiplicative group formed by all non-singular $n \times n$ matrices with entries in K. Indeed $GL_n(K)$ is, to all intents and purposes, just $GL(V)$ when $V = K^n$. In particular it is an affine group. Its highly explicit form has some advantages. For instance the mapping

$$T : GL_n(K) \to GL_n(K)$$

which replaces each matrix by its transpose, is a K-isomorphism of affine sets and an anti-isomorphism of groups.

If K is _infinite_ then, of course, $GL_n(K)$ is connected and of dimension n^2.

Example 4. The affine group $GL_1(K)$, obtained by taking $n = 1$ in the last example, consists of the non-zero elements of K the law of composition being _multiplication._ It is therefore known as the _multiplicative group_ of K. We shall sometimes denote it by G_m.

Suppose that K is _infinite._ Then G_m is connected, its dimension is one, and (2.5.7) shows that

$$K[G_m] = K[X, X^{-1}], \tag{5.3.1}$$

where X denotes an indeterminate.

Consider the direct product

$$G_m \times G_m \times \ldots \times G_m \tag{5.3.2}$$

where there are n factors. Such a group is called a _torus._

Assume once more that K is _infinite._ Then the torus (5.3.2) is connected and its dimension is n. Indeed, with a self-explanatory notation, its coordinate ring may be taken to be

$$K[X_1, X_1^{-1}, X_2, X_2^{-1}, \ldots, X_n, X_n^{-1}],$$

where $X_1, X_2, \ldots, X_n$ are distinct indeterminates.

Example 5. Let us regard K as an affine set and define $\mu : K \times K \to K$ and $j : K \to K$ by $\mu(x, y) = x + y$, $j(x) = -x$. Then $[K, \mu, j, 0]$ is a commutative affine group called the additive group of K. If K is a finite field, then its additive group is a finite group. However if K is infinite the additive group is connected, of dimension one, and its coordinate ring may be taken to be $K[X]$, where X is an indeterminate.

Example 6. Let V be an n-dimensional $(n \geq 1)$ vector space over K. Put $E = \mathrm{End}_K(V)$ and

$$SL(V) = \{g \mid g \in E \text{ and } D(g) = 1\}.$$

Then $SL(V)$ is an affine group. It is known as the special linear group of V.

Assume that K is infinite and let X_{ij} be indeterminates. Note that $SL(V)$ may be regarded either as a closed subset of E or as a closed subgroup of the general linear group $GL(V)$; however in both cases we obtain the same coordinate ring namely

$$K[X_{11}, X_{12}, \ldots, X_{nn}]/I,$$

where I is the ideal formed by the members of

$$K[E] = K[X_{11}, X_{12}, \ldots, X_{nn}]$$

which vanish at all points of $SL(V)$.

Next the determinant $\mathrm{Det}\|X_{ij}\|$ is an irreducible polynomial and therefore $\mathrm{Det}\|X_{ij}\| - 1$ is also irreducible. Let $A = A(X_{22}, \ldots, X_{nn})$ be the cofactor of X_{11} in the matrix $\|X_{ij}\|$ and suppose that $F(X_{11}, \ldots, X_{nn})$ belongs to I. Then, by long division, we obtain a relation

$$A^m F = (\text{Det}\|X_{ij}\| - 1)\, Q + B,$$

where $Q, B \in K[X_{11}, \ldots, X_{nn}]$ and X_{11} does not occur in B.

Let $\alpha_{12}, \alpha_{13}, \ldots, \alpha_{nn}$ belong to K and be such that $A(\alpha_{22}, \ldots, \alpha_{nn}) \neq 0$. We can choose $\alpha_{11} \in K$ so that $\text{Det}\|\alpha_{ij}\| = 1$. Since $F \in I$, the value of $F(\alpha_{11}, \alpha_{12}, \ldots, \alpha_{nn})$ is zero and therefore $B(\alpha_{12}, \ldots, \alpha_{nn}) = 0$. Thus for all choices of $\alpha_{12}, \alpha_{13}, \ldots, \alpha_{nn}$ the polynomial

$$A(X_{22}, \ldots, X_{nn}) B(X_{12}, X_{13}, \ldots, X_{nn})$$

vanishes and therefore, because K is an infinite field, it is the null polynomial. It follows that $B(X_{12}, X_{13}, \ldots, X_{nn})$ is null. Accordingly $A^m F$ belongs to the ideal generated by $\text{Det}\|X_{ij}\| - 1$. But, because $\text{Det}\|X_{ij}\| - 1$ is irreducible, this ideal is prime and now we see that it contains F. It follows that

$$I = (\text{Det}\|X_{ij}\| - 1),$$

and thus we obtain

Theorem 17. Let K be an infinite field and V an n-dimensional $(n \geq 1)$ vector space over K. Then the special linear group $SL(V)$ is connected, its dimension is $n^2 - 1$, and its coordinate ring may be taken to be

$$K[X_{11}, X_{12}, \ldots, X_{nn}]/(\text{Det}\|X_{ij}\| - 1)$$

in the manner explained above.

We mention, in passing, that when working with $GL_n(K)$, rather than with $GL(V)$, the notation for the special linear group, that is the closed subgroup formed by all $n \times n$ matrices with determinant 1, is $SL_n(K)$.

Example 7. Let V be an n-dimensional $(n \geq 1)$ vector space over K and put $E = \text{End}_K(V)$. If $x \in E$ and $v \in V$, then (in what follows) we shall write xv rather than $x(v)$.

Let U, W be subspaces of V with $U \subseteq W$. Put

$$GL(U, W) = \{x \mid x \in GL(V),\ xw - w \in U \text{ all } w \in W\}.$$

For any x in GL(V), the mapping $x : V \to V$ is an automorphism and therefore W and xW have the same dimension. Likewise U and xU have the same dimension. But if $x \in GL(U, W)$, then $xW \subseteq W$ and $xU \subseteq U$. Consequently, $xW = W$ and $xU = U$. A trivial verification now shows that GL(U, W) is a subgroup of GL(V).

Assume that $v \in V$, $\xi \in \mathrm{Hom}_K(V, K) = V^*$, and define

$$\psi_{v, \xi} : E \to K$$

by $\psi_{v, \xi}(x) = \xi(xv)$. Evidently $\psi_{v, \xi}$ belongs to $\mathrm{Hom}_K(E, K)$ and therefore $\psi_{v, \xi} \in K[E]$. Hence the restriction of $\psi_{v, \xi}$ to GL(V) belongs to K[GL(V)].

Suppose that $x \in GL(V)$ and $w \in W$. Then $xw - w \in U$ if and only if $\xi(xw - w) = 0$ for all $\xi \in V^*$ such that $\xi(U) = 0$. But $\xi(xw - w)$ is the value of $\psi_{w, \xi} - \xi(w)$ at x. Consequently GL(U, W) consists of all elements of GL(V) that are common zeros of the functions $\psi_{w, \xi} - \xi(w)$, where w ranges over W and $\xi \in V^*$ satisfies $\xi(U) = 0$. This shows, in particular, that GL(U, W) is a closed subgroup of GL(V) and hence an affine group.

From here on we shall assume that K is an infinite field, w will always denote an element of W, and $\xi \in V^*$ will be assumed to satisfy $\xi(U) = 0$. Select $w_1, w_2, \ldots, w_p$ and $\xi_1, \xi_2, \ldots, \xi_p$ so that

$$\{\psi_{w_\mu, \xi_\mu}\}_{1 \le \mu \le p} \tag{5.3.3}$$

is a base for the K-space spanned by all the functions $\psi_{w, \xi}$ and note that if i denotes the identity of E, then $\psi_{w, \xi}(i) = \xi(w)$. It follows that every $\psi_{w, \xi} - \xi(w)$ is a K-linear combination of the functions

$$\{\xi_{w_\mu, \xi_\mu} - \xi_\mu(w_\mu)\}_{1 \le \mu \le p}. \tag{5.3.4}$$

Accordingly GL(U, W) consists precisely of the elements of GL(V) that are common zeros of (5.3.4).

Now (5.3.3) can be enlarged to a base of $\mathrm{Hom}_K(E, K)$ and, because K is infinite, K[E] is essentially a polynomial ring with the members of the base playing the role of indeterminates.† Accordingly the functions in (5.3.4) generate a prime ideal, P say, in K[E] and P is the intersection of all the K-rational maximal ideals that contain it. Let X be the locus of P. Then X is closed in E, it is irreducible, and $X \cap GL(V) = GL(U, W)$. Thus GL(U, W) is a non-empty open subset of X and therefore, by Chapter 3 Theorem 2, the closure of GL(U, W) in E is X. It follows that if $F \in K[E]$, then F vanishes everywhere on GL(U, W) if and only if $F \in P$. Consequently the ideal of K[GL(V)] that corresponds to GL(U, W) is just the extension of P to K[GL(V)] and, in particular, it is prime. This proves

Theorem 18. _Let_ K _be an infinite field and_ U, W, _where_ $U \subseteq W$, _subspaces of the_ n-_dimensional_ K-_space_ V. _Then (with the above notation)_ GL(U, W) _is a closed, connected subgroup of_ GL(V).

Example 8. Let A be a non-trivial, unitary and associative K-algebra. (It need not be commutative.) Further let A, considered as a vector space over K, have finite dimension. Then A has a natural structure as an affine set. In what follows we use U(A) to denote the group of units of A, that is U(A) consists of those elements which have two-sided inverses.

Suppose that $\alpha \in A$ and let $\lambda_\alpha : A \to A$ be the mapping in which $\lambda_\alpha(a) = \alpha a$. Then λ_α is K-linear and if we put $N(\alpha)$ equal to the determinant of λ_α, we have

$$N(\beta\alpha) = N(\beta)N(\alpha) \qquad (\alpha, \beta \in A)$$

and $N(1_A) = 1_K$. In particular, if α has a right inverse, then $N(\alpha) \neq 0$.

Conversely suppose that $N(\alpha) \neq 0$. Then λ_α is surjective and therefore $\alpha\beta = 1_A$ for some $\beta \in A$. But now $N(\beta) \neq 0$ and hence $\beta\gamma = 1_A$ for some $\gamma \in A$. It follows that $\alpha = \gamma$ and hence that α has a two-sided inverse. Thus

† See Chapter 2 Theorem 19.

$$U(A) = \{\alpha \mid \alpha \in A \text{ and } N(\alpha) \neq 0\}.$$

It is clear that N, considered as a mapping of A into K, belongs to $K[A]$ and therefore we can turn $U(A)$ into an affine set by using the construction described in Example 4 of section (2. 5).

We wish to consider $U(A)$ as an affine group. If K is finite there is no problem so from here on it will be assumed that K is <u>infinite.</u> In this case

$$K[U(A)] = (K[A])[1/N] \tag{5. 3. 5}$$

and $K[A]$ is a polynomial ring in n variables, where n is the dimension of A as a vector space over K.

It is clear that the multiplication mapping

$$U(A) \times U(A) \to U(A)$$

is a K-morphism. Define $j : U(A) \to U(A)$ by $j(\alpha) = \alpha^{-1}$. In order to investigate j we select a base $\alpha_1, \alpha_2, \ldots, \alpha_n$ for A over K. If $\alpha \in A$, then

$$\alpha\alpha_s = \sum_{r=1}^{n} c_{rs}(\alpha)\alpha_r,$$

where $c_{rs} : A \to K$ belongs to $K[A]$ and the determinant of the matrix $\| c_{rs}(\alpha) \|$ is $N(\alpha)$.

Let $1_A = t_1\alpha_1 + t_2\alpha_2 + \ldots + t_n\alpha_n$, where $t_\nu \in K$. Then, for $\alpha \in U(A)$,

$$\alpha^{-1} = d_1(\alpha)\alpha_1 + d_2(\alpha)\alpha_2 + \ldots + d_n(\alpha)\alpha_n$$

provided that

$$\sum_{s=1}^{n} c_{rs}(\alpha)d_s(\alpha) = t_r \qquad (r = 1, 2, \ldots, n).$$

Solving these equations for $d_s(\alpha)$ shows that $d_s : U(A) \to K$ belongs to

$$(K[A])[1/N] = K[U(A)]$$

and thereby establishes that j is a K-morphism. Accordingly we have proved

Theorem 19. *Let K be an infinite field and let the notation be as above. Then $U(A)$ is a connected affine group with $(K[A])[1/N]$ as its coordinate ring. The dimension of $U(A)$ equals the dimension of A as a vector space over K.*

Let V be an n-dimensional $(n \geq 1)$ vector space over K and put $A = \mathrm{End}_K(V)$. Then Theorem 19 applies in this case. It is obvious that $U(A)$ and $GL(V)$ coincide as abstract groups. Also if $f : V \to V$ belongs to A and $D(f)$ denotes the determinant of f, then $N(f) = [D(f)]^n$. This shows that $U(A)$ and $GL(V)$ have the same coordinate ring and therefore they coincide as affine groups.

Example 9. In this example (and the special cases derived from it) K is assumed to be an *infinite field whose characteristic is different from* 2.

Let $n \geq 1$ be an integer and put $G = GL_n(K)$. This is an affine group. If Ω is an extension field of K, then $G^{(\Omega)} = GL_n(\Omega)$ as may be seen by applying Chapter 2 Lemma 14 and Theorem 1 Cor. 1. If A is an $n \times n$ matrix we use $\mathrm{Det}(A)$ and A^T to denote its determinant and transpose respectively. We also use I to denote the $n \times n$ identity matrix.

Suppose that $B \in GL_n(K)$ and put

$$H = \{A \mid A \in M_n(K) \text{ and } A^T BA = B\}.$$

Evidently if $A \in H$ then $\mathrm{Det}(A) \neq 0$. An easy verification shows that H is a closed subgroup of G.

Now put

$$\Lambda = \{P \mid P \in M_n(K) \text{ and } P^T B + BP = 0\}, \qquad (5.3.6)$$

so that Λ is a subspace of the K-space $M_n(K)$, let $P_1, P_2, \ldots, P_q$ be a base for Λ, and set

$$P^* = P_1Z_1 + P_2Z_2 + \dots + P_qZ_q, \tag{5.3.7}$$

where $Z_1, Z_2, \dots, Z_q$ are indeterminates. The entries in P^* are linear forms in $Z_1, Z_2, \dots, Z_q$ and among these we can find q that are linearly independent. Also $\mathrm{Det}(I - P^*) \neq 0$ and $\mathrm{Det}(I - P^{*T}) \neq 0$ as may be seen by putting all the Z_i equal to zero. We may therefore define a matrix A^* by

$$A^* = (I + P^*)(I - P^*)^{-1}. \tag{5.3.8}$$

Then, because $P^{*T}B + P^*B = 0$, we have

$$(I - P^{*T})A^{*T}BA^*(I - P^*) = (I - P^{*T})B(I - P^*)$$

whence

$$A^{*T}BA^* = B. \tag{5.3.9}$$

Again

$$(I - A^*)(I - P^*) = -2P^*$$

and

$$(I + A^*)(I - P^*) = 2I.$$

Since the characteristic of K is not 2, the latter relation shows that $\mathrm{Det}(I + A^*) \neq 0$. Consequently

$$P^* = -(I - A^*)(I + A^*)^{-1}. \tag{5.3.10}$$

It follows, from (5.3.8) and (5.3.10), that the entries in P^* generate, over K, the same field as those in A^* and hence that $K(P^*) = K(A^*)$. (Here P^* is regarded as a generalized point of $M_n(K)$ and A^* as a generalized point of G.) Accordingly

$$K(A^*) = K(P^*) = K(Z_1, Z_2, \dots, Z_q). \tag{5.3.11}$$

The entries in A^* are rational functions in $Z_1, Z_2, \dots, Z_q$ and they can be written with

$$\mathrm{Det}(I - P^*) = \phi(Z_1, Z_2, \ldots, Z_q)$$

(say) as a common denominator. If $Z_1, Z_2, \ldots, Z_q$ are assigned values in K which do not make ϕ vanish, then (5.3.9) shows that A^* is turned into a matrix belonging to H. It follows that if $F \in I_G(H)$, then $F(A^*) = 0$. Consequently, by Chapter 2 Theorem 34, $A^* \in H^{(K(Z))}$ where $K(Z)$ is used as an abbreviation for $K(Z_1, Z_2, \ldots, Z_q)$. -

Let H_0 be the connected component of the identity of H and let $f \in I_H(H_0)$. Then we can find $g \in K[H]$ which vanishes on all the irreducible components of H other than H_0 and is such that $g(I) \neq 0$. Then $fg = 0$ and hence $f(A^*)g(A^*) = 0$. Since I is a specialization of A^* when they are regarded as points of G, the same holds† for H. Consequently $g(A^*) \neq 0$ and therefore $f(A^*) = 0$. Thus, by Chapter 2 Theorem 34, $A^* \in H_0^{(K(Z))}$.

<u>We claim that</u> A^* <u>is a generic point of</u> H_0. To see this first suppose that $A \in H_0$ and $\mathrm{Det}(I + A) \neq 0$. Put

$$P = -(I - A)(I + A)^{-1}$$

so that

$$P^T B = -(I + A^T)^{-1}(I - A^T)B$$

$$BP = -B(I - A)(I + A)^{-1}$$

and therefore $P^T B + BP = 0$. It follows that P can be obtained from P^* by giving $Z_1, Z_2, \ldots, Z_q$ suitable values in K. Next

$$(I - P)(I + A) = 2I$$

whence $\mathrm{Det}(I - P) \neq 0$ and now it follows that

$$A = (I + P)(I - P)^{-1}.$$

Thus in view of (5.3.8) A can be obtained from A^* by giving $Z_1, Z_2, \ldots, Z_q$ suitable values in K and therefore A is a specialization

---

† See Chapter 2, Theorem 34 Cor. 1.

of A^*. Here in the first instance we regard A^* and A as a generalized point and an ordinary point respectively of $GL_n(K)$; but then Chapter 2 Theorem 34 Cor. 1 shows that our conclusion remains valid if we regard them as belonging to H_0.

Finally suppose that $f \in K[H_0]$ and $f(A^*) = 0$. The above remarks show that $f(A)\mathrm{Det}(I + A) = 0$ for all $A \in H_0$. Thus $fh = 0$, where $h \in K[H_0]$ and $h(A) = \mathrm{Det}(I + A)$ when $A \in H_0$. Now $K[H_0]$ is an integral domain and $h \neq 0$ because $h(I) \neq 0$. Accordingly $f = 0$ and we have established our claim that A^* is a generic point of H_0. It follows that $\mathrm{Dim}\, H = \mathrm{Dim}\, H_0$ is equal to the transcendence degree of $K(A^*)$ over K and this is q by (5.3.11). We combine our main conclusions in

Theorem 20. *Let K be an infinite field whose characteristic is different from 2 and let $B \in GL_n(K)$. Then*

$$H = \{A \mid A \in M_n(K) \text{ and } A^T BA = B\}$$

is a closed subgroup of $GL_n(K)$. Put

$$\Lambda = \{P \mid P \in M_n(K) \text{ and } P^T B + BP = 0\}.$$

Then $\mathrm{Dim}\, H$ is equal to the dimension of Λ as a vector space over K. Moreover if A^* is defined as in (5.3.8), then A^* is a generic point of the connected component of the identity of H.

Theorem 21. *Let the assumptions and notation be as in Theorem 20. If the matrix B is symmetric, then $\mathrm{Dim}\, H = n(n-1)/2$. On the other hand, if B is skew symmetric we have $\mathrm{Dim}\, H = n(n+1)/2$.*

Proof. First assume that B is symmetric and let $P \in M_n(K)$. Then $P \in \Lambda$ if and only if $(BP)^T = -BP$. Accordingly Λ consists of all matrices $B^{-1}Q$, where Q is a typical skew symmetric matrix. Thus the vector space Λ has the same dimension as the space of all skew symmetric $n \times n$ matrices and this is $n(n-1)/2$.

Now assume that B is skew symmetric. This time Λ consists of all matrices $B^{-1}S$, where $S \in M_n(K)$ and is symmetric. These symmetric matrices form a vector space of dimension $n(n+1)/2$ and this there fore is the dimension of Λ. The desired results follow by virtue of Theorem 20.

If B is the $n \times n$ identity matrix I_n, then B is non-singular and symmetric. In this case the group H of Theorem 20 is denoted by $O_n(K)$ and called the orthogonal group.

Finally if n is an even integer, say $n = 2m$, we may take B to be the skew symmetric matrix

$$\begin{Vmatrix} O & I_m \\ -I_m & O \end{Vmatrix}$$

This time H is known as the symplectic group. The notation for the symplectic group is $SP_m(K)$.

5.4 K-homomorphisms of affine groups

Throughout section (5.4) G and G' denote affine groups defined over K.

Definition. A mapping $\phi : G \to G'$ is called a 'K-homomorphism' of affine groups if ϕ is a K-morphism of affine sets and $\phi(xy) = \phi(x)\phi(y)$ for all $x, y \in G$.

Note that the identity mapping of G is a K-homomorphism and that K-homomorphisms $\phi : G \to G'$ and $\psi : G' \to G''$ combine to give a K-homomorphism $\psi \circ \phi$ of G into G''.

A K-homomorphism $\phi : G \to G'$ is called a K-isomorphism (of affine groups) if (a) ϕ is a bijection, and (b) $\phi^{-1} : G' \to G$ is a K-homomorphism as well. The affine groups G and G' are said to be K-isomorphic if there exists a K-isomorphism of G on to G'. This relation is reflexive, symmetric and transitive. For example if V is an n-dimensional $(n \geq 1)$ vector space over K, then $GL(V)$ and $GL_n(K)$ are K-isomorphic affine groups.

Let $\phi : G \to G'$ be a K-homomorphism of affine groups and put $N = \operatorname{Ker} \phi$. Then N is a closed subgroup of G because ϕ is continuous and N is the inverse image of the identity of G'. For instance we have a K-homomorphism $GL(V) \to GL_1(K)$ in which each f in $GL(V)$ is mapped into its determinant. In this case the kernel is $SL(V)$.

Theorem 22. *Let $\phi : G \to G'$ be a K-homomorphism of affine groups and suppose that K is algebraically closed. Then $\phi(G)$ is a closed subgroup of G'.*

Proof. It is enough to show that $\phi(G)$ is closed in G'. Let G_0 be the component of the identity of G and assume that we can show that $\phi(G_0)$ is closed in G'. Since, by Theorem 10, $\phi(G)$ is the union of a finite number of cosets of $\phi(G_0)$, the desired result will then follow. We shall therefore assume that G is connected and hence *irreducible.*

Put $H = \overline{\phi(G)}$. Then H is a connected, closed subgroup of G' an ϕ induces an almost surjective K-homomorphism $\tilde{\phi} : G \to H$. Theorem 3 of Chapter 3 now shows that $\phi(G) = \tilde{\phi}(G)$ contains a non-empty subset U which is open in H. However, by Theorem 14 Cor., $H = UU \subseteq \phi(G)$. The theorem follows.

Corollary. *Let $\phi : G \to G'$ be a K-homomorphism of affine group If ϕ is almost surjective and K is algebraically closed, then ϕ is surjective.*

For the general situation, where K is not necessarily algebraicall closed, we have

Theorem 23. *Let $\phi : G \to G'$ be an almost surjective K-homomor phism of affine groups. Further let G_0 respectively G'_0 be the connect component of the identity of G respectively G'. Then $\overline{\phi(G_0)} = G'_0$.*

Proof. By Theorem 10, the index of G_0 in G is finite. Let $x_1, x_2, \ldots, x_m$ be representatives of the different cosets. Then

$$\phi(G) = \bigcup_{i=1}^{m} \phi(x_i)\phi(G_0)$$

and therefore

$$G' = \overline{\phi(G)} = \bigcup_{i=1}^{m} \phi(x_i)\overline{\phi(G_0)}.$$

Thus $\overline{\phi(G_0)}$ is a closed subgroup of G' whose index in G' is finite. Consequently, by Theorem 11, $G'_0 \subseteq \overline{\phi(G_0)}$. On the other hand $\phi(G_0)$ is connected and contains the identity of G'. Accordingly $\overline{\phi(G_0)} \subseteq G'_0$ and the theorem is proved.

Corollary. _Let the assumptions be as in the theorem and suppose that_ K _is algebraically closed. Then_ $\phi(G_0) = G'_0$.

This now follows from Theorem 22.

Before proceeding to the next theorem we note that if $\phi : G \to G'$ is a K-homomorphism of affine groups and L is an extension field of K, then $\phi^{(L)} : G^{(L)} \to G'^{(L)}$ is an L-homomorphism of affine groups. (This is easily seen by using Lemma 1.) Also if ϕ is almost surjective, then so too is $\phi^{(L)}$. The latter assertion follows from (2.10.10).

Theorem 24. _Let_ $\phi : G \to G'$ _be an almost surjective_ K-_homomorphism of affine groups and put_ $N = \operatorname{Ker} \phi$. _Then_ $\operatorname{Dim} N \le \operatorname{Dim} G - \operatorname{Dim} G'$. _If_ K _is algebraically closed, then_ $\operatorname{Dim} N = \operatorname{Dim} G - \operatorname{Dim} G'$.

Proof. We begin by considering the case where K is algebraically closed. Let G_0 respectively G'_0 be the connected component of the identity of G respectively G'. Then $\operatorname{Dim} G_0 = \operatorname{Dim} G$, $\operatorname{Dim} G'_0 = \operatorname{Dim} G'$, and by Theorem 23 Cor., ϕ induces a surjective K-homomorphism $G_0 \to G'_0$ whose kernel is $N \cap G_0 = N_1$ (say). Since the index of G_0 in G is finite, the index of N_1 in N is finite as well. It follows that $\operatorname{Dim} N_1 = \operatorname{Dim} N$ and we have reduced the problem (for the algebraically closed case) to the situation where G and G' are connected and hence irreducible.

By Chapter 3 Theorem 37 Cor., there exists $y \in G'$ such that $\phi^{-1}(\{y\})$ has a component whose dimension is $\operatorname{Dim} G - \operatorname{Dim} G'$. Let x belong to this component. Then $\phi^{-1}(\{y\}) = xN$. But N and xN are K-isomorphic and every component of N has dimension equal to Dim N by virtue of the corollary to Theorem 10. This shows that $\operatorname{Dim} N = \operatorname{Dim} G - \operatorname{Dim} G'$ and we have dealt with the case where K is algebraically closed.

Now let K be an arbitrary field and let L be its algebraic closure.

We know that ϕ extends to an almost surjective L-homomorphism $\phi^{(L)} : G^{(L)} \to G'^{(L)}$. Put $N^* = \operatorname{Ker} \phi^{(L)}$. Then $N \subseteq N^*$ and

$$\operatorname{Dim} N^* = \operatorname{Dim} G^{(L)} - \operatorname{Dim} G'^{(L)} = \operatorname{Dim} G - \operatorname{Dim} G'.$$

By Chapter 2 Theorem 30, $N^{(L)} \subseteq N^*$. Consequently

$$\operatorname{Dim} N = \operatorname{Dim} N^{(L)} \leq \operatorname{Dim} N^*$$

and with this the proof is complete.

5.5 K-morphic actions on an affine set

Let G be an affine group and V a non-empty affine set where both G and V are defined over K. A left action of G on V is a mapping

$$\alpha : G \times V \to V,$$

where

(i) $\alpha(\sigma, \alpha(\tau, v)) = \alpha(\sigma\tau, v)$ for all $\sigma, \tau \in G$ and $v \in V$;
(ii) $\alpha(e, v) = v$, where $e = e_G$ and $v \in V$.

When we have such an action we normally write σv in place of $\alpha(\sigma, v)$. In this notation (i) and (ii) become $\sigma(\tau v) = (\sigma\tau)v$ and $ev = v$ respectively.

Suppose that α is a left action of G on V and that α is also a K-morphism of affine sets. We then say that α is a K-morphic left action of G on V or, less formally, that G acts morphically on the left of V.

A K-morphic right action of G on V is a K-morphism $V \times G \to V$ with analogous properties. Thus in a right action $(v\sigma)\tau = v(\sigma\tau)$ and $ve = e$. Of course results concerning right and left actions tend to come in pairs. We shall develop the theory mainly in terms of left actions.

Suppose that we have a K-morphic left action of G on V. The mapping

$$\lambda_\sigma : V \to V \tag{5.5.1}$$

in which $\lambda_\sigma(v) = \sigma v$ is a K-automorphism of V with $\lambda_{\sigma^{-1}}$ as its inverse. It therefore induces an automorphism

$$\lambda^*_\sigma : K[V] \to K[V] \tag{5.5.2}$$

of K-algebras. It will be convenient to put

$$\lambda^*_\sigma(f) = f^\sigma. \tag{5.5.3}$$

Accordingly we have the following formulae for a left action:

$$\left.\begin{aligned} f^\sigma(v) &= f(\sigma v), \\ (f_1 + f_2)^\sigma &= f_1^\sigma + f_2^\sigma, \\ (f_1 f_2)^\sigma &= f_1^\sigma f_2^\sigma, \\ (f^\sigma)^\tau &= f^{\sigma\tau}, \\ k^\sigma &= k \text{ for } k \in K. \end{aligned}\right\} \tag{5.5.4}$$

Similar remarks apply to K-morphic right actions. The corresponding formulae for a right action are

$$\left.\begin{aligned} f^\sigma(v) &= f(v\sigma), \\ (f_1 + f_2)^\sigma &= f_1^\sigma + f_2^\sigma, \\ (f_1 f_2)^\sigma &= f_1^\sigma f_2^\sigma, \\ (f^\sigma)^\tau &= f^{\tau\sigma}, \\ k^\sigma &= k \text{ for } k \in K. \end{aligned}\right\} \tag{5.5.5}$$

If G is an affine group, then the multiplication mapping

$$\mu : G \times G \to G$$

is a K-morphism and it defines both a left and a right action of G on itself. These are known respectively as the regular left action and regular right action. Contrary to what was said above, when we consider a regular action of G on itself it will usually be the regular right action that will concern us. The reason for this will appear later.

Theorem 25. *Let the affine group G act morphically on the left of the affine set V and let $f \in K[V]$. Then the K-space spanned by all f^σ, where f is fixed and σ varies in G, has finite dimension.*

This is essentially a generalization of part of Theorem 6. Since the proof involves no new ideas we shall not give details.

At this point it is convenient to note that if $\alpha : G \times V \to V$ is a K-morphic left action of G on V and L is an extension field of K, then $\alpha^{(L)} : G^{(L)} \times V^{(L)} \to V^{(L)}$ is an L-morphic left action of $G^{(L)}$ on $V^{(L)}$.

Once again let the affine group G act morphically on the left of the affine set V, and for $v \in V$ put

$$\mathrm{Stab}_G(v) = \{\sigma | \sigma \in G \text{ and } \sigma v = v\}.$$

Obviously $\mathrm{Stab}_G(v)$ is a subgroup of the abstract group G. Now $\sigma \mapsto \sigma v$ is a K-morphism $G \to V$ and we have a second K-morphism of G into V in which every element of G is mapped into the given element v. An application of Lemma 1 at once shows that $\mathrm{Stab}_G(v)$ *is a closed subgroup of* G.

Suppose that $v_1, v_2 \in V$ and let us write $v_1 \sim v_2$ if $v_2 = \sigma v_1$ for some $\sigma \in G$. This is an equivalence relation and the equivalence classes are known as *orbits.* The orbit that contains the element v is Gv.

Let X be a subset of V. We say that X is *stable* under G if $\sigma X \subseteq X$ whenever $\sigma \in G$. When this is the case we have $\sigma X = X$ for all $\sigma \in G$, and the orbit of any $x \in X$ is contained in X.

Suppose that X is stable under G. Then its closure $\overline{X}$, in V, is also stable under G. It follows that $\overline{X} \backslash X$ is a stable set as well. Hence if N is an orbit, then $\overline{N} \backslash N$ is stable under G.

Theorem 26. *Let the connected affine group G act morphically on the left of the affine set V, let N be an orbit, and let K be algebraically closed. Then N is open in its closure $\overline{N}$. If the orbit N is such that $\mathrm{Dim}\, \overline{N}$ is as small as possible (for the different orbits of G in V), then $N = \overline{N}$ (that is N is a closed orbit).*

Proof. Let $v \in N$ and consider the K-morphism $G \to \overline{N}$ in which $\sigma \mapsto \sigma v$. The morphism is almost surjective and its image is N. Consequently, by Chapter 3 Theorem 33, there exists a non-empty subset T, of $\overline{N}$, which is open in $\overline{N}$ and contained in N. Next

$$N = \bigcup_{\sigma \in G} \sigma T .$$

Also $\overline{N}$ is stable under G and therefore each σT is open in $\overline{N}$. It follows that N is open in $\overline{N}$.

Now assume that $w \in \overline{N} \backslash N$ and put $N_0 = Gw$. Then $N_0 \subseteq \overline{N} \backslash N$ whence $\overline{N}_0 \subseteq \overline{N} \backslash N$ because N is open in $\overline{N}$. Thus $\overline{N}_0 \subseteq \overline{N}$, $\overline{N}_0 \neq \overline{N}$. Moreover $\overline{N}$ is irreducible because G is irreducible. Accordingly, by Chapter 3 Theorem 22, $\operatorname{Dim} \overline{N}_0 < \operatorname{Dim} \overline{N}$. Hence if $\operatorname{Dim} \overline{N}$ is minimal, then $N = \overline{N}$.

Corollary. <u>Let</u> G <u>be an affine group (not necessarily connected) which acts morphically on the left of the affine set</u> V. <u>Further let</u> N <u>be one of the orbits and suppose that</u> $\operatorname{Dim} \overline{N}$ <u>is minimal (in the sense of the theorem). Then</u> $N = \overline{N}$.

Proof. First suppose that N is <u>any</u> orbit. It is clear that G_0 acts morphically on V. Choose $\sigma_0 = e, \sigma_1, \ldots, \sigma_m$ to represent the different cosets of G_0 in G and let $v \in N$. Put $N_0 = G_0 v$ and

$$N_i = \sigma_i N_0 = \sigma_i G_0 v = G_0(\sigma_i v).$$

Then $N_0, N_1, \ldots, N_m$ are orbits of G_0,

$$N = N_0 \cup N_1 \cup \ldots \cup N_m,$$

and

$$\overline{N} = \overline{N}_0 \cup \overline{N}_1 \cup \ldots \cup \overline{N}_m.$$

Next each $\overline{N}_i$ is irreducible and $\overline{N}_i = \sigma_i(\overline{N}_0)$. Thus the $\overline{N}_i$ are pairwise K-isomorphic. It follows that $\operatorname{Dim} \overline{N} = \operatorname{Dim} \overline{N}_i$ for all i. Accordingly when $\operatorname{Dim} \overline{N}$ has the smallest possible value the same holds for the closure of the G_0-orbit N_i. Consequently $\overline{N}_i = N_i$, by the theorem, and hence $\overline{N} = N$.

5.6 G-modules

Let G be an arbitrary group and V a vector space over K. Suppose that a mapping

$$\alpha : G \times V \to V$$

has been given and let us denote the image of (σ, v), where $\sigma \in G$ and $v \in V$, by σv. If now

(i) $\sigma(v_1 + v_2) = \sigma v_1 + \sigma v_2$ for $v_1, v_2 \in V$ and $\sigma \in G$;

(ii) $\sigma(kv) = k(\sigma v)$ for $k \in K$, $\sigma \in G$, and $v \in V$;

(iii) $\sigma_1(\sigma_2 v) = (\sigma_1 \sigma_2)v$ for $\sigma_1, \sigma_2 \in G$ and $v \in V$;

(iv) $ev = v$ when $e = e_G$ and $v \in V$;

then we say that V is a left (G, K)-module. If we are in a situation where K is being kept fixed, then we often say simply that V is a left G-module. This applies particularly when G is an affine group. In this case it is always to be understood that K is the ground field over which G is defined.

The notion of a right (G, K)-module is obtained by making the obvious modifications to the above definition. Thus a right G-module has to do with a mapping $V \times G \to V$ in which the image of (v, σ) is denoted by $v\sigma$, $(v\sigma)\tau = v(\sigma\tau)$, and so on.

We can express the above ideas in ring-theoretic terms. To be explicit, let KG denote the group ring of G with coefficients in K. Thus KG is the K-space which has the members of G as a base; and if

$$\sum_{\sigma \in G} k_\sigma \sigma \quad \text{and} \quad \sum_{\tau \in G} k'_\tau \tau$$

are two elements of KG, then their product is given by

$$\left(\sum_{\sigma \in G} k_\sigma \sigma\right)\left(\sum_{\tau \in G} k'_\tau \tau\right) = \sum\sum k_\sigma k'_\tau (\sigma\tau).$$

It is clear that the notion of a left respectively right (G, K)-module is essentially the same as a left respectively right KG-module as these latter terms are understood in the theory of rings and modules.

Suppose now that V and W are left (say) G-modules. A mapping $f : V \to W$ is called a G-homomorphism if it is a linear mapping of K-spaces and $f(\sigma v) = \sigma f(v)$ for all $\sigma \in G$ and $v \in V$. If in addition to being a G-homomorphism f is also bijective, then f is called a G-isomorphism. In terms of the group ring KG, a G-homomorphism respectively G-isomorphism is the same as a homomorphism respectively isomorphism of KG-modules.

Now suppose that G is an affine group defined over K and that V is a (G, K)-module, where the dimension of V over K is finite. (For definiteness we shall suppose that V is a left G-module.) If $\sigma \in G$, then the mapping

$$\lambda_\sigma : V \to V \tag{5.6.1}$$

in which $\lambda_\sigma(v) = \sigma v$, belongs to $\mathrm{End}_K(V)$. Consequently, there is a mapping

$$\lambda : G \to \mathrm{End}_K(V), \tag{5.6.2}$$

where $\lambda(\sigma) = \lambda_\sigma$. (A similar situation is obtained if V is a right G-module.)

Definition. If the G-module V has finite dimension over K, then V is called a 'rational' G-module if (5.6.2) is a K-morphism of affine sets.

Once again let V be a finite-dimensional G-module and let $u_1, u_2, \ldots, u_n$ be a base for V over K. Then

$$\lambda_\sigma(u_j) = \sum_{i=1}^{n} \lambda_{ij}(\sigma)u_i,$$

where the λ_{ij} are mappings of G into K.

Lemma 3. Let G be an affine group and V a finite-dimensional left G-module. Further let the notation be as described above. In these circumstances the following statements are equivalent:

(a) V is a rational G-module;

(b) all the functions λ_{ij} belong to K[G]:

(c) the mapping $\alpha : G \times V \to V$, where $\alpha(\sigma, v) = \sigma v$, is K-morphic.

Remark. Naturally a similar result holds concerning right G-modules.

Proof. We may assume that K is infinite, for if K is finite, then (a), (b) and (c) all hold trivially.

Assume (a) and define $\omega_{\mu\nu} : \mathrm{End}_K(V) \to K$ for $1 \le \mu, \nu \le n$ by $\omega_{\mu\nu}(f) = a_{\mu\nu}$, where $\|a_{ij}\|$ is the matrix of f with respect to $u_1, u_2, \dots, u_n$. Then $\omega_{\mu\nu} \in K[\mathrm{End}_K(V)]$ and so $\lambda_{\mu\nu} = \omega_{\mu\nu} \circ \lambda \in K[G]$ because $\lambda : G \to \mathrm{End}_K(V)$ is a K-morphism by hypothesis. Thus (a) implies (b).

Assume (b). There exist $\gamma_1, \gamma_2, \dots, \gamma_n$ in K[V] such that if $v \in V$, then

$$v = \gamma_1(v)u_1 + \gamma_2(v)u_2 + \dots + \gamma_n(v)u_n.$$

Hence for $f \in K[V]$ we have

$$\begin{aligned} f(\sigma v) &= f(\lambda_\sigma(v)) \\ &= f\Big(\sum_j \gamma_j(v)\lambda_\sigma(u_j)\Big) \\ &= f\Big(\sum_i\sum_j \gamma_j(v)\lambda_{ij}(\sigma)u_i\Big) \\ &= q\Big(\sum_j \lambda_{1j}(\sigma)\gamma_j(v), \sum_j \lambda_{2j}(\sigma)\gamma_j(v), \dots, \sum_j \lambda_{nj}(\sigma)\gamma_j(v)\Big), \end{aligned}$$

where $q(X_1, X_2, \dots, X_n)$ is a certain polynomial with coefficients in K. Hence if we use the notation introduced in (2. 7. 2), then

$$\omega \circ \alpha = q\Big(\sum_j \lambda_{ij} \vee \gamma_j, \sum_j \lambda_{2j} \vee \gamma_j, \dots, \sum_j \lambda_{nj} \vee \gamma_j\Big)$$

and this belongs to $K[G \times V]$. Thus (b) implies (c).

Finally assume (c). Since $\sigma u_j = \sum_{i=1}^{n} \lambda_{ij}(\sigma)u_i$, it follows that

$$\lambda_{ij}(\sigma) = \gamma_i(\sigma u_j) = (\gamma_i \circ \alpha)(\sigma, u_j),$$

where γ_i is defined as in the last paragraph. But $\gamma_i \circ \alpha \in K[G \times V]$

and thus we see that $\lambda_{ij} \in K[V]$ for all i and j. Let $\chi \in K[\mathrm{End}_K(V)]$. There is a polynomial $p(X_{11}, X_{12}, \ldots, X_{nn})$, with coefficients in K, such that if $f \in \mathrm{End}_K(V)$ and has matrix $\|a_{ij}\|$, then $\chi(f) = p(a_{11}, a_{12}, \ldots, a_{nn})$. Accordingly

$$(\chi \circ \lambda)(\sigma) = p(\lambda_{11}(\sigma), \lambda_{12}(\sigma), \ldots, \lambda_{nn}(\sigma))$$

and therefore

$$\chi \circ \lambda = p(\lambda_{11}, \lambda_{12}, \ldots, \lambda_{nn}) \in K[G].$$

Thus λ is a K-morphism and we have shown that (c) implies (a). This establishes the lemma.

As before let V be a left G-module of finite dimension. The mapping $\lambda_\sigma : V \to V$ is not just an endomorphism but an automorphism with inverse $\lambda_{\sigma^{-1}}$. Thus we arrive at a homomorphism

$$\lambda_0 : G \to GL(V) \tag{5.6.3}$$

of abstract groups, where $\lambda_0(\sigma) = \lambda_\sigma$. (In the case of a right G-module we obtain an <u>anti-homomorphism.</u>) If (5.6.3) is a K-homomorphism of affine groups, then we say that we have a <u>rational representation</u> of G by means of automorphisms of V.

Corollary. <u>Let G be an affine group and V a finite-dimensional left G-module. Then V is a rational G-module if and only if (5.6.3) is a rational representation of</u> G.

Proof. It is clear that we may assume that K is infinite. Next the natural mapping $GL(V) \to \mathrm{End}_K(V)$ is a K-morphism so half the corollary is clear. Now assume that V is a rational G-module and let $\chi \in K[GL(V)]$. If f with matrix $\|a_{ij}\|$ is in GL(V), then

$$\chi(f) = p(a_{11}, \ldots, a_{nn})/[\mathrm{Det}(f)]^h,$$

where $p(X_{11}, \ldots, X_{nn})$ belongs to the polynomial ring $K[X_{11}, \ldots, X_{nn}]$ and $h > 0$ is an integer. Hence

$$(\chi \circ \lambda_0)(\sigma) = p(\lambda_{11}(\sigma), \ldots, \lambda_{nn}(\sigma))/[\mathrm{Det}(\lambda_0(\sigma))]^h .$$

Thus

$$\chi \circ \lambda_0 = p(\lambda_{11}, \ldots, \lambda_{nn})/[\mathrm{Det} \circ \lambda_0]^h$$

and we have to show that this belongs to K[G]. By the lemma, all the λ_{ij} are members of K[G]. Consequently $p(\lambda_{11}, \ldots, \lambda_{nn}) \in K[G]$. Consider the mapping $\sigma \mapsto \mathrm{Det}(\lambda_0(\sigma))$. Since this is just $\mathrm{Det} \circ \lambda$ it belongs to K[G]. It follows that $\sigma \mapsto [\mathrm{Det}(\lambda_0(\sigma^{-1}))]^h$ also belongs to K[G]. But

$$[\mathrm{Det}(\lambda_0(\sigma^{-1}))]^h = [\mathrm{Det}\ \lambda_0(\sigma)]^{-h}$$

because $\lambda_0(\sigma^{-1})$ is the inverse of $\lambda_0(\sigma)$. Accordingly $[\mathrm{Det} \circ \lambda_0]^{-h}$ belongs to K[G] and the corollary follows.

Observe that this result shows that the notion of a finite-dimensional rational G-module is equivalent to that of a rational representation of G by automorphisms of a finite-dimensional K-space.

Example 1. Let V be a finite-dimensional vector space over K. For $f \in GL(V)$ and $v \in V$, put $fv = f(v)$. This turns V into a rational left GL(V)-module.

Example 2. Let G be an affine group and let G act on itself by means of the regular right action. For $f \in K[G]$ and $\sigma \in G$ put $\sigma f = f^\sigma$. Then, by (5. 5. 5), the mapping $f \mapsto \sigma f$ is K-linear, $ef = f$ and

$$\sigma(\tau f) = (f^\tau)^\sigma = f^{\sigma\tau} = (\sigma\tau)f.$$

Thus K[G] has become a left (G, K)-module.

Let us keep f fixed and denote by N the K-subspace of K[G] which is generated by the elements f^σ $(\sigma \in G)$. By Theorem 25, the dimension of N as a vector space over K is finite. Also $\sigma N = N$ for all $\sigma \in G$ and, moreover, $f \in N$. Thus N is a left (G, K)-module and it contains f. The next example shows that N is a rational G-module.

Example 3. Let G be an affine group and suppose that M is a subspace of the K-space K[G]. We assume that the dimension of M (over K) is finite and that $\sigma M = M$ for all $\sigma \in G$. Here $K[G]$ is regarded as a left G-module as in Example 2.

Lemma 4. With the above assumptions M is a rational G-module.

Proof. Let $f_1, f_2, \ldots, f_n$ be a base for M over K. Then

$$\sigma f_j = f_j^\sigma = \sum_{i=1}^{n} \lambda_{ij}(\sigma) f_i$$

for $\sigma \in G$, where $\lambda_{ij} : G \to K$. Accordingly for $y \in G$,

$$f_j(y\sigma) = \sum_{i=1}^{n} \lambda_{ij}(\sigma) f_i(y).$$

If y is kept fixed, then the mapping $G \to K$ in which $\sigma \mapsto f_j(y\sigma)$ belongs to $K[G]$. Next, by Chapter 2 Theorem 13, we can find $y_1, y_2, \ldots, y_n$ in G so that the determinant $|f_i(y_k)|$ is non-zero. Hence, by solving the equations

$$f_j(y_k\sigma) = \sum_{i=1}^{n} \lambda_{ij}(\sigma) f_i(y_k) \qquad (k = 1, 2, \ldots, n),$$

we find that $\lambda_{ij}(\sigma)$ is a certain linear combination (with coefficients in K) of $f_j(y_1\sigma), f_j(y_2\sigma), \ldots, f_j(y_n\sigma)$. This shows that $\lambda_{ij} \in K[G]$ and therefore the desired result follows from Lemma 3.

We shall now give an important application of the ideas contained in Examples 2 and 3.

Let G be an affine group, choose $f_1, f_2, \ldots, f_s$ so that $K[G] = K[f_1, f_2, \ldots, f_s]$, and let M be the subspace of the K-space $K[G]$ generated by the set f_i^σ $(1 \le i \le s, \sigma \in G)$. Theorem 25 shows that the dimension of M is finite and evidently $\sigma M = M$ for all $\sigma \in G$. Accordingly, by Lemma 4, M is a rational G-module. Let us select $g_1, g_2, \ldots, g_n$ so that they form a base of M over K. Note that $K[G] = K[g_1, g_2, \ldots, g_n]$ and, by Lemma 3 Cor., we have a K-homomorphism

$$\rho : G \to GL(M) \tag{5.6.4}$$

of affine groups. Suppose that $\sigma \in \operatorname{Ker} \rho$. Then $g_i^\sigma = g_i$ and thus right translation by means of σ induces the identity automorphism of the K-algebra $K[G]$. Accordingly $\sigma = e$ and therefore $\text{Ker}\,\rho = \{e\}$.

From here on we shall assume that K is algebraically closed. By Theorem 22, $\rho(G)$ is a closed subgroup of $GL(M)$. Hence ρ induces a bijective K-morphism

$$\bar{\rho} : G \to \rho(G) \tag{5.6.5}$$

of affine groups.

To ρ there corresponds a homomorphism

$$\rho^* : K[GL(M)] \to K[G]$$

of K-algebras. We claim that ρ^* is surjective. To see this define $\rho_{ij} : G \to K$ by

$$(\rho(\sigma))(g_j) = g_j^\sigma = \sum_{i=1}^n \rho_{ij}(\sigma) g_i.$$

Then

$$g_j(\sigma) = g_j^\sigma(e) = \sum_{i=1}^n \rho_{ij}(\sigma) g_i(e).$$

Next let the polynomial $F_j(X_{11}, \ldots, X_{nn})$ be given by

$$F_j(X_{11}, \ldots, X_{nn}) = \sum_{i=1}^n X_{ij} g_i(e)$$

and regard it as belonging to $K[GL(M)]$. Then

$$\begin{aligned}(F_j \circ \rho)(\sigma) &= F_j(\rho(\sigma)) \\ &= F_j(\rho_{11}(\sigma), \ldots, \rho_{nn}(\sigma)) \\ &= \sum_{i=1}^n \rho_{ij}(\sigma) g_i(e) \\ &= g_j(\sigma).\end{aligned}$$

Consequently $\rho^*(F_j) = g_j$ and our claim follows.

Next we have homomorphisms

$$G \to \rho(G) \to GL(M)$$

where the first is $\bar{\rho}$ and the second is an inclusion mapping. These give rise to homomorphisms

$$K[GL(M)] \to K[\rho(G)] \to K[G]$$

of K-algebras. We have just seen that

$$\rho^* : K[GL(M)] \to K[G]$$

is surjective and now it follows that

$$\bar{\rho}^* : K[\rho(G)] \to K[G]$$

is surjective as well. Let $g \in K[G]$. Then g is the image of some $\psi \in K[\rho(G)]$, that is to say $\psi \circ \bar{\rho} = g$. But $\bar{\rho}$ is bijective. Consequently

$$g \circ \bar{\rho}^{-1} = \psi \in K[\rho(G)].$$

Accordingly

$$\bar{\rho}^{-1} : \rho(G) \to G$$

is a K-morphism and therefore

$$\bar{\rho} : G \to \rho(G)$$

is a K-isomorphism of affine groups. Thus we have proved the following striking result.

Theorem 27. *Let* G *be an affine group defined over an algebraically closed ground field* K. *Then there exists a finite-dimensional* K-*space* M *such that the group* G *is* K-*isomorphic to a closed subgroup of* GL(M).

5.7 General rational G-modules

If G is an affine group and V is a G-module, then already in section (5.6) we have explained what is meant by saying that V is a *rational* G-module in the restricted situation where the dimension of V (over the ground field) is *finite.* This concept will now be extended. First, however, we shall introduce some general terminology.

Suppose that G is an abstract group and V a left (G, K)-module. Further let U be a subset of V. Then U is said to be a (G, K)-*submodule* of V provided

(a) U is a K-subspace of V;

(b) $\sigma u \in U$ for all $\sigma \in G$ and $u \in U$.

Note that if KG denotes the group ring of G, then a (G, K)-submodule of V is the same as a KG-submodule in the sense in which this term is used in the theory of rings and modules. (Naturally similar definitions and observations apply to right (G, K)-modules.) Often we omit any explicit reference to K and speak of a G-module and its G-submodules. In what follows a G-module is said to be *finite-dimensional* when its dimension as a K-space is finite.

Lemma 5. *Let G be an affine group and V a finite-dimensional G-module. If now V is a rational G-module, then all its G-submodules are rational.*

This follows immediately from Lemma 3.

Let G be an affine group, V a G-module of arbitrary dimension, and U a G-submodule of V. Then U is a G-module and if it happens to be finite-dimensional, then it could be a rational G-module in the sense of section (5.6).

Lemma 6. *Let $U_1, U_2, \ldots, U_m$ be finite-dimensional rational G submodules of the G-module V. Then $U_1 + U_2 + \ldots + U_m$ is a finite-dimensional rational G-submodule of V.*

Proof. We may assume that $m = 2$. Choose a base for $U_1 \cap U_2$ and extend it so as to obtain bases for U_1 and U_2. In this way we obtain a special base for $U_1 + U_2$. The desired result follows by applying Lemma 3 to this base.

Once again let G be an affine group and V a G-module.

General definition. We say that V is a 'rational' G-module if every element of V is contained in a finite-dimensional rational G-submodule.

Note that if V itself is finite-dimensional, then the new definition agrees with the old. This is so in view of Lemma 6.

Theorem 28. _Let G be an affine group and V a rational G-module. Then every G-module of V is a rational G-module._

This follows from Lemma 5 and the definition of a general rational G-module.

We have already seen that, when G is an affine group, the coordinate ring K[G] can be regarded as a left G-module by making use of the regular right action of G on itself. In this situation we have $\sigma f = f^{\sigma}$, when $f \in K[G]$ and $\sigma \in G$.

Theorem 29. _Let G be an affine group and let K[G] be considered as a left G-module in the manner described above. Then K[G] is a rational G-module._

This follows from Examples 2 and 3 in section (5.6).

Definition. A G-module V is said to be 'simple' if $V \neq 0$ and the only G-submodules of V are the zero submodule and V itself.

Theorem 30. _Let G be an affine group and V a rational G-module. If V is a simple G-module, then the dimension of V (over the ground field) is finite._

This is clear from the definitions.

Theorem 31. _Let G be an affine group and $V \neq 0$ a rational G-module. Then V contains a simple G-submodule._

Proof. We can find a G-submodule $U \neq 0$ whose dimension is finite. If we arrange that the dimension is as small as possible, then U will be a simple G-submodule.

5.8 Linearly reductive affine groups

Let G be an abstract group.

Definition. A (G, K)-module V is said to be 'completely reducible' if given any (G, K)-submodule U there exists a second (G, K)-submodule U' such that $V = U \oplus U'$ this being a direct sum of vector spaces.

Evidently every (G, K)-submodule of a completely reducible (G, K)-module is itself completely reducible.

Lemma 7. Let V be a (G, K)-module. Then the following statements are equivalent:

(a) V is completely reducible;

(b) V is a sum of simple (G, K)-modules;

(c) V is a direct sum of simple (G, K)-modules.

Proof. If one restates the lemma in terms of modules over the group ring KG, then it becomes a special case of a familiar result in the theory of semi-simple modules over a ring.†

Definition. An affine group G is said to be 'linearly reductive' if every rational G-module‡ is completely reducible.

Theorem 32. Let G be an affine group and suppose that every finite-dimensional rational G-module is completely reducible. Then every rational G-module is completely reducible, that is to say G is linearly reductive.

Proof. Let V be a rational G-module. Then V is a sum of finite-dimensional rational G-submodules and each of these is a sum of simple G-submodules. Thus V itself is a sum of simple G-submodules and therefore it is completely reducible.

Let G be an affine group and let $U_1, U_2, \ldots, U_m$ be G-modules. Their direct sum

$$U = U_1 \oplus U_2 \oplus \ldots \oplus U_m$$

---

† See, for example, [(3) Theorem 11, p. 61].

‡ For the remainder of this chapter all G-modules will be understood to be *left* G-modules.

has a natural structure as a G-module. Evidently if each U_i is a rational G-module, then U is a rational G-module. Again if each U_i is a completely reducible G-module, then the same holds for U.

Suppose that V is a G-module and that $d > 0$ is an integer. As is customary we write

$$V^d = V \oplus V \oplus \dots \oplus V,$$

where there are d summands.

At this point we make a fresh start. Let G be an affine group and V an n-dimensional, rational G-module. By Theorem 29, K[G] has a natural structure as a rational G-module. Let $F \in V^* = \mathrm{Hom}_K(V, K)$, let $v \in V$ and define

$$\overline{F}(v) : G \to K$$

by

$$(\overline{F}(v))(\sigma) = F(\sigma v).$$

Since V is a rational G-module, it follows readily that $\overline{F}(v) \in K[G]$. Evidently

$$\overline{F} : V \to K[G]$$

is K-linear; also if $\tau \in G$, then

$$\begin{aligned}(\overline{F}(\tau v))(\sigma) &= F(\sigma\tau v) \\ &= (\overline{F}(v))(\sigma\tau) \\ &= (\overline{F}(v))^{\tau}(\sigma)\end{aligned}$$

and therefore $\overline{F}(\tau v) = \tau\overline{F}(v)$. Consequently $\overline{F} : V \to K[G]$ is a G-homomorphism.

Let $F_1, F_2, \dots, F_n$ be a base for V^* over K and define a mapping $V \to (K[G])^n$ by

$$v \mapsto (\overline{F}_1(v), \overline{F}_2(v), \dots, \overline{F}_n(v)).$$

This is both a G-homomorphism and an injection. We have therefore proved

Theorem 33. _Let_ G _be an affine group and let_ K[G] _be regarded as a rational left_ G-_module as in Theorem 29. If now_ V _is an_ n-_dimensional rational_ G-_module, then there is a_ G-_submodule of_ $(K[G])^n$ _which is_ G-_isomorphic to_ V.

Theorem 34. _Let_ G _be an affine group and let_ K[G] _be regarded as a rational_ G-_module as in Theorem 29. Then_ G _is linearly reductive if and only if_ K[G] _is completely reducible._

This follows from Theorems 32 and 33.

Let us return to the proof of Theorem 33. The argument used show that if $V \neq 0$, then, for some i, the G-homomorphism $\overline{F}_i : V \to K[G]$ is not null. Hence if V is a _simple_ rational G-module, then we can find $F \in V^*$ such that the G-homomorphism $\overline{F} : V \to K[G]$ is not null and henc an injection. This observation establishes

Theorem 35. _Let_ G _be an affine group and let_ K[G] _be regarded as a rational_ G-_module as in Theorem 29. Let_ V _be a simple, rational_ G-_module. Then_ V _is_ G-_isomorphic to a_ G-_submodule of_ K[G].

If G is a finite group, then every finite-dimensional G-module is rational, and therefore all G-modules are rational. The classical theory of finite groups shows that if K has characteristic zero or the characteristic of K does not divide the order of G, then every G-module is completely reducible. This is because KG is a semi-simple ring by virtue of Maschke's Theorem.†

It is not easy to give other examples of linearly reductive groups without making use of the theory of Lie algebras, but an exception arises in the case of _tori._ These groups are of considerable interest and we shall take this opportunity to discuss some of their properties.

In section (5. 3) we defined a torus to be a finite direct product

$$G = G_m \times G_m \times \ldots \times G_m \quad (s \text{ factors}), \tag{5.8.1}$$

where $G_m = GL_1(K)$ is the multiplicative group of K. It is convenient to extend this definition and describe as a torus any affine group which is

† See, for example, [(10) §125, p. 193].

K-isomorphic to one of these groups. We shall also regard a trivial group (that is a group with no elements apart from its identity) as a torus. This amounts to admitting the possibility that, in (5.8.1), the integer s may be zero.

Let us suppose that K is infinite. Then the coordinate ring of the group G in (5.8.1) is

$$K[X_1, X_1^{-1}, X_2, X_2^{-1}, \ldots, X_s, X_s^{-1}],$$

where $X_1, X_2, \ldots, X_s$ are indeterminates. Assume that $\nu_1, \nu_2, \ldots, \nu_s$ are integers (not necessarily positive), let $c \in K$ and put $F = cX_1^{\nu_1}X_2^{\nu_2} \ldots X_s^{\nu_s}$. If now $\sigma = (\sigma_1, \sigma_2, \ldots, \sigma_s)$, where $\sigma_i \in G_m$, then it is easy to check that $F^\sigma = \sigma_1^{\nu_1}\sigma_2^{\nu_2} \ldots \sigma_s^{\nu_s}F$. Hence the subspace of $K[G]$ generated by $X_1^{\nu_1}X_2^{\nu_2} \ldots X_s^{\nu_s}$ is a G-submodule and, since it is one-dimensional, it is simple. Note that we have shown that $K[G]$ is a sum of one-dimensional G-submodules.

Next suppose that H is a closed subgroup of G. If V is a G-module, then it has a natural structure as an H-module; and if it is a finite-dimensional rational G-module, then it is also a finite-dimensional rational H-module. Hence any rational G-module can be regarded as a rational H-module.

Theorem 36. *Let G be a torus defined over an infinite field K and let H be a closed subgroup of G. Then K[H], considered as a rational H-module as in Theorem 29, is a direct sum of one-dimensional (and hence simple) H-submodules. Consequently H is linearly reductive.*

Proof. We have a surjective homomorphism $K[G] \to K[H]$ of K-algebras and each of $K[G]$ and $K[H]$ is a rational H-module. An easy verification shows that $K[G] \to K[H]$ is a homomorphism of H-modules. Now $K[G]$ is a sum of one-dimensional G-submodules and these submodules are also H-submodules. It follows that $K[H]$ is a sum of one-dimensional H-submodules. Finally this sum can be refined to give a direct sum by omitting suitably selected summands.

Corollary. *Let G, K and H be as in Theorem 36 and let V be a rational H-module. Then V is a direct sum of one-dimensional H-submodules.*

Proof. Let S be a rational, simple H-module. In view of Theorem 36 it will suffice to show that S has dimension one. By Theorem 35, S is H-isomorphic to a submodule of the completely reducible H-module K[H] and therefore it is a homomorphic image of K[H]. Accordingly the simple H-module S is a sum of one-dimensional H-submodules and therefore it too must be one-dimensional.

Let us take these ideas a little further.

Theorem 37. *Let H be an affine group defined over an algebraically closed field K. Then the following statements are equivalent:*

(i) *every rational H-module is a direct sum of one-dimensional H-submodules;*

(ii) *every finite-dimensional rational H-module is a direct sum of one-dimensional H-submodules;*

(iii) *H is a closed subgroup of a torus.*

Proof. The equivalence of (i) and (ii) is clear. Also the corollary to Theorem 36 shows that (iii) implies (i).

Assume (ii). We shall deduce (iii) and in view of Theorem 27 we may suppose that H is a closed subgroup of GL(V), where V is some finite-dimensional vector space over K. Now V is a rational GL(V)-module and therefore it is a rational H-module. Consequently, since we are assuming (ii), we can find a base $e_1, e_2, \ldots, e_n$ of V with the property that each Ke_i is an H-submodule of V.

With the usual notation

$$K[GL(V)] = K[X_{11}, X_{12}, \ldots, X_{nn}]\left[\frac{1}{\operatorname{Det}\|X_{ij}\|}\right]$$

so if we put

$$G = \{f \mid f \in GL(V) \text{ and } X_{ij}(f) = 0 \text{ whenever } i \neq j\},$$

then G is a closed subgroup of GL(V) containing H. However the mapp-

$$f \mapsto (X_{11}(f), X_{22}(f), \ldots, X_{nn}(f))$$

is a K-isomorphism of G on to

$$G_m \times G_m \times \ldots \times G_m \qquad \text{(n factors)}$$

and therefore we see that H is a closed subgroup of a torus. The proof is now complete.

5.9 Characters and semi-invariants

Let G be an affine group defined over K.

Definition. A 'rational character' of G is a K-homomorphism $\chi : G \to GL_1(K)$ of affine groups.

Thus a rational character of G is a function $\chi \in K[G]$ which gives rise to a group-homomorphism of G into the group formed by the non-zero elements of the ground field.

In order to illustrate one way in which rational characters can occur naturally we introduce a further definition. To this end let M be a rational left G-module and suppose that $m \in M$.

Definition. We say that m is an 'eigenvector' of G if (i) $m \neq 0$, and (ii) $\sigma m \in Km$ for all $\sigma \in G$.

Suppose that m is an eigenvector of G and define $\chi : G \to K$ by $\sigma m = \chi(\sigma)m$. Then, because M is a rational G-module, $\chi \in K[G]$ and, since $\sigma^{-1}(\sigma m) = m$, we have $\chi(\sigma) \neq 0$. In fact χ is a rational character of G. Thus a rational G-module gives rise to rational characters via its eigenvectors.

Lemma 8. _Let the rational G-module M have finite dimension. Then the number of rational characters, of G, to which M gives rise is finite._

Proof. Choose linearly independent eigenvectors $m_1, m_2, \ldots, m_s$ with s as large as possible and let χ_i be the rational character associated with m_i. Now let m be any other eigenvector, in M, and χ the corresponding rational character. In this situation we have a relation $m = a_1 m_1 + a_2 m_2 + \ldots + a_s m_s$, where $a_i \in K$ and we may suppose that $a_1 \neq 0$. Let $\sigma \in G$ and apply σ to the relation. This yields $a_1 \chi(\sigma) = a_1 \chi_1(\sigma)$ whence $\chi = \chi_1$. This proves the lemma.

Let G be an affine group over K, H a closed subgroup of G, and M a rational left G-module. Then M is also a rational H-module.

Lemma 9. *Suppose that* G *is connected and that* H *is a closed, normal subgroup of* G. *Assume that* m, *in* M, *is an eigenvector of* H *and let* χ *be the corresponding rational character of* H. *Then* $\chi(\sigma\lambda\sigma^{-1}) = \chi(\lambda)$ *for all* $\sigma \in G$ *and* $\lambda \in H$.

Proof. Without loss of generality we may assume that the dimensi of M, when considered as a vector space over K, is finite. Next for $\sigma \in G$ and $\lambda \in H$ we have $\sigma\lambda\sigma^{-1}(\sigma m) = \chi(\lambda)(\sigma m)$. Consequently σm is an eigenvector of H and if we denote the corresponding rational character of H by χ_σ, then $\chi_\sigma(\sigma\lambda\sigma^{-1}) = \chi(\lambda)$. In particular $\chi_e = \chi$.

By Lemma 8, M gives rise to only finitely many different rational characters of H. Let these be $\chi = \chi_1, \chi_2, \ldots, \chi_r$. Then, for fixed $\lambda \in H$ and fixed j $(1 \leq j \leq r)$,

$$\{\sigma | \sigma \in G, \ \chi(\sigma^{-1}\lambda\sigma) = \chi_j(\lambda)\}$$

is a closed subset of G, and therefore

$$\{\sigma | \sigma \in G, \ \chi_\sigma = \chi_j\} = V_j$$

say is also closed in G. But

$$G = V_1 \cup V_2 \cup \ldots \cup V_r,$$

the union is disjoint and G is connected. Thus one of the sets V_i is G itself and the others are empty. It follows that $\chi_\sigma = \chi$ for all $\sigma \in G$ and now the lemma follows.

Let G be an affine group over K. By Theorem 29, K[G] has a natural structure as a left G-module. Let H be a closed subgroup of G.

Definition. An element of K[G] which is an eigenvector of H is called a 'semi-invariant' of H.

Suppose that $f \in K[G]$, $f \neq 0$ is a semi-invariant of H and let $\chi : H \to K$ be the corresponding rational character. We say, in these circumstances, that the semi-invariant f has weight χ. Note that $f^\lambda = \chi(\lambda)f$ for all $\lambda \in H$ and that $\chi(\lambda)$ is the restriction to H of a member of K[G].

Theorem 38. *Let G be an affine group defined over K and let H be a closed subgroup of G. Then we can find semi-invariants $p_1, p_2, \ldots, p_r$ of H such that*

(a) *$p_1, p_2, \ldots, p_r$ all have the same weight;*

(b) *H consists precisely of those $\sigma \in G$ such that $p_\mu^\sigma \subseteq Kp_\mu$ for $\mu = 1, 2, \ldots, r$.*

In particular the closed subgroups of G are determined by their semi-invariants.

Proof. We may suppose that $H \neq G$. Let $g_1, g_2, \ldots, g_t$ generate the ideal $I_G(H)$ and let M be the K-subspace of K[G] spanned by all g_i^σ $(\sigma \in G, 1 \le i \le t)$. By Theorem 25, the dimension of M is finite; moreover M is a rational left G-module. On the other hand, if $f \in I_G(H)$ and $\lambda \in H$, then also $f^\lambda \in I_G(H)$. Consequently if we put

$$N = I_G(H) \cap M,$$

then N is not only a finite-dimensional K subspace of M but also a rational H-module.

Suppose that $\sigma \in G$ and $N^\sigma \subseteq N$. Since $g_1, g_2, \ldots, g_t$ are in N, the ideal NK[G], generated by N, is none other than $I_G(H)$. This shows that if $f \in I_G(H)$, then also $f^\sigma \in I_G(H)$ and therefore

$$f(\sigma) = f^\sigma(e) = 0.$$

It follows that

$$H = \{\tau \mid \tau \in G \text{ and } N^\tau \subseteq N\}.$$

Let the dimension of N (as a vector space over K) be d and consider the exterior power $\Lambda^d M$. This has an obvious structure as a left G-module and as such it is rational; moreover $\Lambda^d M$ contains the one-dimensional K-space $\Lambda^d N$ and this subspace is a rational H-module.

Choose $u \in \Lambda^d N$ so that $u \neq 0$. If $\lambda \in H$ then certainly $\lambda u \in Ku$. Now suppose that $\sigma \in G$ and $\sigma u \in Ku = \Lambda^d N$. <u>We claim that</u> $\sigma \in H$. For let us take a base of N and extend it to a base of M. Then consider the effect of σ on the elements which make up the base of N. Our claim follows from the fact that if a $d \times q$ $(d \leq q)$ K-matrix has exactly one non-singular $d \times d$ submatrix, then the $d \times q$ matrix has only d non-zero columns. Thus we see that

$$H = \{\sigma | \sigma \in G \quad \text{and} \quad \sigma u \in Ku\}.$$

Construct a base $u = e_0, e_1, e_2, \ldots, e_s$ of $\Lambda^d M$ over K and for $\sigma \in G$ write

$$\sigma e_j = \sum_{i=0}^{s} f_{ij}(\sigma) e_i,$$

where $f_{ij} : G \to K$. In fact $f_{ij} \in K[G]$. Next for all $\sigma, \tau \in G$

$$f_{ij}(\sigma\tau) = \sum_{\mu=0}^{s} f_{i\mu}(\sigma) f_{\mu j}(\tau).$$

Also $\tau \in H$ if and only if $\tau e_0 \in Ke_0$, that is $\tau \in H$ if and only if $f_{\mu 0}(\tau) = 0$ for $\mu = 1, 2, \ldots, s$.

Suppose now that $\sigma \in G$ and $\lambda \in H$. Then

$$f_{i0}^{\lambda}(\sigma) = f_{i0}(\sigma\lambda) = f_{00}(\lambda) f_{i0}(\sigma)$$

and so we see that $f_{i0}^{\lambda} = f_{00}(\lambda) f_{i0}$. Accordingly the non-zero members of $\{f_{10}, f_{20}, \ldots, f_{s0}\}$ are semi-invariants of H and they all have the same weight namely the restriction of f_{00} to H.

Finally assume that $\sigma \in G$ and $f_{i0}^{\sigma} \in Kf_{i0}$ for $i = 1, 2, \ldots, s$. Then, for $1 \leq i \leq s$, $f_{i0}(e) = 0$ because $e \in H$ and therefore $f_{i0}(\sigma) = f_{i0}^{\sigma}(e) = 0$. It follows that $\sigma \in H$. Thus we can complete the proof by taking $p_1, p_2, \ldots, p_r$ to be the non-zero members of the set $\{f_{10}, f_{20}, \ldots, f_{s0}\}$.

We shall now interrupt the main discussion in order to make some fairly general observations concerning operations with rational modules over an affine group. These will assist us when we come to apply Theorem 38.

To this end let G be an affine group defined over K, and let M, N be finite-dimensional rational left G-modules. Then $M \otimes_K N$ has a natural structure as a rational left G-module in which

$$\sigma(m \otimes n) = \sigma m \otimes \sigma n.$$

Here, of course, $m \in M$, $n \in N$ and $\sigma \in G$.

Now put $M^* = \mathrm{Hom}_K(M, K)$. For $f \in M^*$ and $\sigma \in G$ define σf to be the mapping $m \mapsto f(\sigma^{-1}m)$ of M into K. Then $\sigma f \in M^*$ and it is easily checked that M^* has become a left (G, K)-module. In order to examine this structure more closely, let $e_1, e_2, \ldots, e_t$ be a base of M over K and $e_1^*, e_2^*, \ldots, e_t^*$ the dual base. Suppose that

$$\sigma e_j = \sum_{i=1}^{t} f_{ij}(\sigma)e_i,$$

where $f_{ij} : G \to K$. Then the f_{ij} are in $K[G]$ and

$$\sigma e_\nu^* = \sum_{p=1}^{t} f_{\nu p}(\sigma^{-1})e_p^*$$

and we see, in particular, that M^* is a <u>rational</u> left G-module. Note that if it happens that for a certain $\sigma \in G$ we have $\sigma e_j = \alpha e_j$ (where $\alpha \in K$) for $j = 1, 2, \ldots, t$, then $\sigma e_\nu^* = \alpha^{-1}e_\nu^*$ for $\nu = 1, 2, \ldots, t$.

The considerations of the preceding two paragraphs enable us to regard $M \otimes_K M^*$ as a rational left G-module. Let $\sigma \in G$ and suppose that when it acts on M the effect is the same as multiplying all the elements of M by a non-zero scalar α. Then the endomorphism of $M \otimes_K M^*$ induced by σ is the identity endomorphism. Now the converse also holds. For suppose that $\tau \in G$ and

$$\tau(e_j \otimes e_\nu^*) = e_j \otimes e_\nu^*$$

for all j and ν. Then $\tau e_j = \alpha_j e_j$ and $\tau e_\nu^* = \beta_\nu e_\nu^*$, where α_j, $\beta_\nu \in K$

and $\alpha_j\beta_\nu = 1$ for all j and ν. It follows that $\alpha_1 = \alpha_2 = \ldots = \alpha_t$ and therefore, when τ operates on M, the effect is the same as that produced when we multiply by this scalar.

Lemma 10. *Let G be an affine group over K and let H be a closed, normal subgroup of G. Assume that either (i) G is connected or (ii) G is commutative. Then there exists a finite-dimensional rational left G-module N such that the corresponding rational representation $G \to GL(N)$ has kernel H.*

Proof. Choose semi-invariants $p_1, p_2, \ldots, p_r$ of H as in Theorem 38 and let χ be their common weight. Denote by M the K-subspace of $K[G]$ spanned by all p_i^σ, where $\sigma \in G$ and $1 \leq i \leq r$. Then M is a finite-dimensional rational left G-module, and p_i^σ is a semi-invariant of H whose weight is the mapping $\lambda \mapsto \chi(\sigma^{-1}\lambda\sigma)$. But $\chi(\sigma^{-1}\lambda\sigma) = \chi(\lambda)$ by Lemma 9 if G is connected and the relation holds trivially if G is commutative. Consequently every non-zero element of M is a semi-invariant of H of weight χ.

Let $\sigma \in G$. By Theorem 38, $\sigma \in H$ if and only if σ acts on M like a scalar multiplier. But this happens if and only if σ induces the identity mapping on $M \otimes_K M^*$. It follows that the rational representation

$$G \to GL(M \otimes_K M^*)$$

has kernel H.

Theorem 39. *Let G be a torus over an infinite ground field K and H a closed subgroup of G. Then there exist rational characters $\chi_1, \chi_2, \ldots, \chi_s$ of G such that H is the intersection of their kernels.*

Proof. By Lemma 10, there exists a rational representation $G \to GL(V)$ of G whose kernel is H. (Here V is a finite-dimensional, left G-module.) By Theorem 36 Cor., we can find a base $e_1, e_2, \ldots, e_m$ of V that is composed of eigenvectors of G. Let χ_i be the rational character of G that corresponds to e_i. Then an element σ, of G, belongs to H if and only if $\chi_i(\sigma) = 1_K$ for $i = 1, 2, \ldots, m$. The theorem follows.

We insert here a few remarks about character groups. These will enable us to exploit the theorem just proved.

Let G be an affine group and χ_1, χ_2 rational characters of G. We define their sum† as characters by

$$(\chi_1 + \chi_2)(\sigma) = \chi_1(\sigma)\chi_2(\sigma).$$

The effect of this is to turn the set of rational characters into an additive abelian group. This group will be denoted by $\hat{G}$. Note that every member of $\hat{G}$ is a unit of $K[G]$.

Now suppose that K is an infinite field and consider the torus

$$G = G_m \otimes G_m \otimes \ldots \otimes G_m \quad \text{(n factors)}$$

so that, with the usual notation,

$$K[G] = K[X_1, X_1^{-1}, X_2, X_2^{-1}, \ldots, X_n, X_n^{-1}].$$

It is easily verified that, in this case, the rational characters are the functions $X_1^{\nu_1}X_2^{\nu_2} \ldots X_n^{\nu_n}$, where the ν_i are integers; and so it follows that (with respect to the addition introduced above) the rational characters form a free abelian group on n generators. Indeed $X_1, X_2, \ldots, X_n$ is one of the bases of this group. Let $\chi_1, \chi_2, \ldots, \chi_n$ be an arbitrary base. We obtain an automorphism of G by means of the mapping

$$\sigma \mapsto (\chi_1(\sigma), \chi_2(\sigma), \ldots, \chi_n(\sigma)).$$

Observe that this is not just an automorphism of abstract groups. Both the mapping and its inverse are K-morphisms of affine sets.

Theorem 40. Let G be a torus, H a closed subgroup of G, and suppose that the ground field K is infinite. Then H is K-isomorphic to an affine group of the form

$$\Gamma_1 \times \Gamma_2 \times \ldots \times \Gamma_q \times G_m \times G_m \times \ldots \times G_m,$$

where Γ_i denotes a finite subgroup of G_m.

† This addition corresponds to multiplication in $K[G]$.

Proof. In what follows χ denotes a typical rational character of G. Put

$$\tilde{H} = \{\chi \mid \chi(\lambda) = 1_K \text{ for all } \lambda \in H\}.$$

This is a subgroup of $\hat{G}$ and, by Theorem 39,

$$H = \{\sigma \mid \sigma \in G \text{ and } \chi(\sigma) = 1_K \text{ for all } \chi \in \tilde{H}\}.$$

But $\hat{G}$ is a free abelian group. We can therefore find a base $\chi_1, \chi_2, \ldots, \chi_n$ for $\hat{G}$ and positive integers $t_1, t_2, \ldots, t_q$ $(q \leq n)$ so that $t_1\chi_1, t_2\chi_2, \ldots, t_q\chi_q$ is a base for $\tilde{H}$. Thus an element σ, of G, belongs to H if and only if $\chi_i(\sigma)^{t_i} = 1_K$ for $i = 1, 2, \ldots, q$.

Consider the K-automorphism of

$$G = G_m \times G_m \times \ldots \times G_m \quad (n \text{ factors})$$

in which $\sigma \to (\chi_1(\sigma), \chi_2(\sigma), \ldots, \chi_n(\sigma))$. The image of H consists of all elements $(y_1, y_2, \ldots, y_n)$ such that $y_i^{t_i} = 1_K$ for $i = 1, 2, \ldots, q$. Hence if

$$\Gamma_i = \{y \mid y \in K \text{ and } y^{t_i} = 1_K\},$$

then H and the group

$$\Gamma_1 \times \Gamma_2 \times \ldots \times \Gamma_q \times G_m \times G_m \times \ldots \times G_m \qquad (n \text{ factors})$$

are K-isomorphic.

Corollary 1. _Let_ H _be a closed subgroup of a torus and suppose that the ground field_ K _is algebraically closed. Then the elements of_ H _that are of finite order are everywhere dense in_ H.

Proof. The theorem allows us to assume that

$$H = \Gamma_1 \times \ldots \times \Gamma_q \times G_m \times \ldots \times G_m,$$

where Γ_i is a finite subgroup of G_m. Consider the elements of finite order in G_m. There are infinitely many of them and so they are everywhere dense in G_m. The corollary now follows by applying Theorem 28 Cor. of Chapter 2.

The next corollary holds for an arbitrary ground field.

Corollary 2. Let H be a closed connected subgroup of a torus. Then H itself is (K-isomorphic to) a torus.

Proof. If K is finite, then H is a finite group and therefore, because it is connected, it has no elements other than its identity element. We may therefore suppose that K is infinite in which case our theorem allows us to assume that

$$H = \Gamma_1 \times \dots \times \Gamma_q \times G_m \times \dots \times G_m.$$

where Γ_i is a finite subgroup of G_m. However in this case Γ_i is connected because it is a continuous image of the connected group H. Accordingly the groups Γ_i are all trivial and therefore H is a direct product $G_m \times G_m \times \dots \times G_m$, i.e. it is a torus.

5.10 Linearly reductive groups and invariant theory

The notion of a linearly reductive group is especially useful in invariant theory, and in this section we shall endeavour to explain why this is so. In what follows G always denotes an affine group defined over the ground field K, and we recall our earlier agreement that all G-modules which occur are understood to be left G-modules unless there is an explicit statement to the contrary.

Suppose that M is a rational G-module. Then, by Theorem 28, every G-submodule of M is also rational. Put

$$M^G = \{x \mid x \in M \text{ and } \sigma x = x \text{ for all } \sigma \in G\}. \tag{5.10.1}$$

This is a G-submodule of M. We shall call it the submodule of *G-invariant* elements of M.

Theorem 41. *Suppose that G is linearly reductive and let M be a rational G-module. Then the set of G-submodules N, of M, such that*

$N^G = 0$ has a (unique) member, M_G say, which contains all the others. This satisfies $M = M^G \oplus M_G$ and it is, moreover, the unique complementary G-submodule of M^G in M.

Proof. It is clear that the G-submodules N form a non-empty inductive system with respect to inclusion. Let N' be a maximal member of this system.

Let N be any G-submodule of M such that $N^G = 0$. We claim that $N \subseteq N'$. (Note that when this is established we shall be able to define M_G.) For suppose that $N \not\subseteq N'$. Then, because G is linearly reductive, N contains a simple G-submodule P such that $P \not\subseteq N'$. Accordingly $P \cap N' = 0$ and therefore the sum $N' + P$ is direct. We also have $P^G = 0$ because $P \subseteq N$. It follows that

$$(N' \oplus P)^G = N'^G \oplus P^G = 0.$$

However this contradicts the maximality of N' and thereby establishes our claim.

Let Q be a simple G-submodule of M. Then either $Q^G = Q$ or $Q^G = 0$, that is either $Q \subseteq M^G$ or $Q \subseteq M_G$. Thus, in any event, $Q \subseteq M^G + M_G$. But M is a sum of simple G-submodules. Consequently $M \subseteq M^G + M_G$ and therefore, since $M^G \cap M_G = 0$, we have $M = M^G \oplus M_G$.

Finally suppose that $M = M^G \oplus T$, where T is some G-submodule of M. Then $T \cap M^G = 0$ and therefore $T^G = 0$. Thus $T \subseteq M_G$ and now we see that T must be equal to M_G. This completes the proof.

Suppose for the moment that G is linearly reductive and let M be a rational G-module. We have just seen that

$$M = M^G \oplus M_G. \tag{5.10.2}$$

Denote by

$$P_M : M \rightarrow M^G \tag{5.10.3}$$

the associated projection of M on to M^G. Note that P_M is a homomorphism of G-modules and that

$$\operatorname{Ker} P_M = M_G. \tag{5.10.4}$$

This projection is known as the Reynold's operator of M.

Let X and Y be rational G-modules and $\omega : X \to Y$ a homomorphism of G-modules.

Lemma 11. *Let G be linearly reductive. Then $\omega(X^G) \subseteq Y^G$ and $\omega(X_G) \subseteq Y_G$. Hence if P_X and P_Y are the Reynold's operators of X and Y respectively, then*

$$\omega(P_X(x)) = P_Y(\omega(x))$$

for all $x \in X$.

Proof. It is obvious that $\omega(X^G) \subseteq Y^G$. Now let Q be a simple G-submodule of X_G. Then either $\omega(Q)$ is G-isomorphic to Q or $\omega(Q) = 0$. In any event $(\omega(Q))^G = 0$ and therefore $\omega(Q) \subseteq Y_G$. But X_G is the sum of its simple G-submodules and so we see that $\omega(X_G) \subseteq Y_G$. The rest of the lemma follows.

Corollary. *Let G be linearly reductive, M a rational G-module, and N a G-submodule of M. Then the restriction of P_M to N coincides with P_N.*

We now turn our attention from modules to algebras. Let R be a non-trivial, unitary, associative and commutative K-algebra. Suppose further that the affine group G acts on R in such a way that (i) R is a rational G-module, and (ii) $\sigma(fg) = (\sigma f)(\sigma g)$ for all f, g in R and σ in G. Then if we fix the element σ, of G, the mapping $f \mapsto \sigma f$ is an automorphism of the K-algebra R. In view of this we shall describe the whole situation by saying that G *acts rationally on* R *by means of K-algebra automorphisms.* If we have this situation, then R^G is a K-subalgebra of R. We refer to it as the K-algebra of G-invariant elements.

Lemma 12. *Let the situation be as described above and let G be linearly reductive. Then R^G and R_G are R^G-submodules of R, and for $f \in R$ and $g \in R^G$ we have $P_R(gf) = gP_R(f)$.*

Proof. It is clear that R^G is an R^G-submodule of R. Suppose now that $g \in R^G$. Then the mapping $R \to R$ in which $f \mapsto gf$ is a G-homomorphism and therefore $gR_G \subseteq R_G$ by Lemma 11. Accordingly R_G is an R^G-submodule of R. The final assertion of Lemma 12 is now trivial.

Corollary 1. _Let the linearly reductive group_ G _act rationally on_ R _by means of K-algebra automorphisms and let_ $\mathfrak{U}$ _be an ideal of_ R^G. _Then_ $\mathfrak{U}R \cap R^G = \mathfrak{U}$.

Proof. Since $R = R^G \oplus R_G$ and R_G is an R^G-module, we have

$$\mathfrak{U}R = \mathfrak{U}R^G \oplus \mathfrak{U}R_G = \mathfrak{U} \oplus \mathfrak{U}R_G$$

and $\mathfrak{U}R_G \subseteq R_G$. Consequently $\mathfrak{U}R \cap R^G = \mathfrak{U}$ as required.

Corollary 2. _Let the linearly reductive group_ G _act rationally on_ R _by means of K-algebra automorphisms. Further let_ $\{\mathfrak{B}_j\}_{j \in J}$ _be a family of ideals of_ R, _where each_ $\mathfrak{B}_j$ _is also a G-submodule of_ R. _Then_

$$\sum_{j \in J} (\mathfrak{B}_j \cap R^G) = (\sum_{j \in J} \mathfrak{B}_j) \cap R^G.$$

Proof. We need only prove that the right hand side is contained in the left hand side since the converse is trivial. Suppose therefore that f belongs to $(\Sigma \mathfrak{B}_j) \cap R^G$. Then

$$f = \sum_{j \in J} f_j,$$

where $f_j \in \mathfrak{B}_j$ and almost all the f_j are zero. Next

$$f = P_R(f) = \sum_{j \in J} P_R(f_j).$$

However $P_R(f_j) \in \mathfrak{B}_j \cap R^G$ because, by Lemma 11 Cor., the restriction of P_R to $\mathfrak{B}_j$ is the Reynolds operator of $\mathfrak{B}_j$. Accordingly f belongs to $\Sigma(\mathfrak{B}_j \cap R^G)$ and the lemma follows.

The next lemma provides a general result from the theory of graded rings. It will be needed in the proof of our main result concerning algebras of G-invariant elements.

Suppose that R is a commutative ring with an identity element (not necessarily a K-algebra) which is graded by the non-negative integers. Thus

$$R = R_0 \oplus R_1 \oplus R_2 \oplus \dots ,$$

where R_μ is a subgroup of the additive group of R and $R_\mu R_\nu \subseteq R_{\mu+\nu}$ for all $\mu \geq 0$ and $\nu \geq 0$. The elements of R_μ are said to be _homogeneous of degree_ μ and any ideal which can be generated by homogeneous elements is called a _homogeneous ideal._ For example

$$R_+ = R_1 \oplus R_2 \oplus R_3 \oplus \dots$$

is a homogeneous ideal of R. Should it happen that the ideal R_+ is finitely generated, then we can find a finite system of generators that is composed of homogeneous elements of positive degree. Note that the identity element of R is necessarily homogeneous of degree zero, that R_0 is a subring of R, and that each R_μ is an R_0-module.

Lemma 13. _Suppose that_ R _is a commutative graded ring with an identity element. If_ R_+ _is a finitely generated ideal of_ R, _then it is possible to find a finite set_ $\{u_1, u_2, \dots, u_p\}$ _of elements of_ R _such that_ $R = R_0[u_1, u_2, \dots, u_p]$.

Proof. Choose $f_1, f_2, \dots, f_s$, where f_i is homogeneous of degree $d_i > 0$, so that

$$R_+ = Rf_1 + Rf_2 + \dots + Rf_s.$$

Put $k = \max\{d_1, d_2, \dots, d_s\}$. If now $n \geq k$ and $f \in R_n$, then

$$f \in R_{n-d_1}f_1 + R_{n-d_2}f_2 + \dots + R_{n-d_s}f_s.$$

Consider the subring S, of R, that is generated by $R_0, R_1, \dots, R_k$ and the elements $f_1, f_2, \dots, f_s$. It is easily verified, using induction, that $R_\mu \subseteq S$ for every $\mu \geq 0$. Consequently $S = R$. The theorem will therefore follow if we show that each R_μ is a finitely generated R_0-module.

Assume that this is not so and choose $p \geq 0$ <u>as small as possible</u> so that R_p is not a finitely generated R_0-module. Then $p \geq 1$. Also

$$R_{p+1} \oplus R_{p+2} \oplus R_{p+3} \oplus \dots$$

is a homogeneous ideal of R. If we factor out this ideal, then we obtain a new graded ring

$$R' = R_0 \oplus \dots \oplus R_p \oplus 0 \oplus 0 \oplus \dots$$

and, because R_+ is mapped on to R'_+, R'_+ must be a finitely generated ideal of R'.

Now put

$$\mathfrak{U}' = R_1 R_{p-1} + R_2 R_{p-2} + \dots + R_{p-1} R_1.$$

This is a homogeneous ideal of R' and, by the choice of p, it is finitely generated as an R_0-module. Let us factor out $\mathfrak{U}'$. This leads to a new graded ring R" whose grading is given by

$$R'' = R_0 \oplus \dots \oplus R_{p-1} \oplus (R_p/\mathfrak{U}') \oplus 0 \oplus 0 \oplus \dots$$

because $\mathfrak{U}' \subseteq R_p$. Also this construction ensures that R''_+ is a finitely-generated R"-ideal.

Let us take a finite homogeneous system of generators for the ideal R''_+ and pick out those that have degree p. These will generate $R''_p = R_p/\mathfrak{U}'$ as an R_0-module because $R''_a R''_b = 0$ if $a > 0$, $b > 0$ and $a + b = p$. Hence $R_p/\mathfrak{U}'$ is a finitely generated R_0-module and therefore R_p is a finitely generated R_0-module. This contradiction completes the proof.

Corollary. <u>Let</u> R <u>be a commutative ring with an identity element which is graded by the non-negative integers. If now</u> R_0 <u>is a Noetherian ring, then the following two statements are equivalent:</u>

(a) R <u>is a Noetherian ring;</u>

(b) $R = R_0[u_1, u_2, \dots, u_n]$ <u>for suitable elements</u> $u_1, u_2, \dots,$ <u>in</u> R.

Proof. Assume (a). Then R_+ is a finitely generated ideal of R and therefore (b) follows by the lemma. The converse holds by virtue of Hilbert's Basis Theorem.

We come now to the main result of this section.

Theorem 42. Let G be a linearly reductive affine group and let R be a unitary associative and commutative K-algebra. Further let G act rationally on R by means of K-algebra automorphisms. If now R is a finitely generated K-algebra, then R^G is also a finitely generated K-algebra.

Proof. By Lemma 12 Cor., $\mathfrak{U}R \cap R^G = \mathfrak{U}$ for every ideal $\mathfrak{U}$ of R^G. Since R is Noetherian, this observation shows that R^G is Noetherian as well.

Choose $v_1, v_2, \ldots, v_s$ in R so that $R = K[v_1, v_2, \ldots, v_s]$ and then choose a finite-dimensional G-submodule M, of R, so that $v_i \in M$ for all i. Next select a K-base $u_1, u_2, \ldots, u_n$ for M. We now have $R = K[u_1, u_2, \ldots, u_n]$ and $Ku_1 + Ku_2 + \ldots + Ku_n$ is a rational G-module.

Let $X_1, X_2, \ldots, X_n$ be indeterminates. There is an isomorphism

$$KX_1 + KX_2 + \ldots + KX_n \approx Ku_1 + Ku_2 + \ldots + Ku_n$$

of K-spaces in which X_i is matched with u_i. We use this to turn $KX_1 + KX_2 + \ldots + KX_n$ into a rational G-module in such a way that our isomorphism of K-spaces becomes an isomorphism of G-modules.

Put $S = K[X_1, X_2, \ldots, X_n]$. Then S is a graded ring with grading

$$S = S_0 \oplus S_1 \oplus S_2 \oplus S_3 \oplus \ldots,$$

where $S_0 = K$ and S_ν consists of all forms of degree ν. Each $\sigma \in G$ determines a K-algebra automorphism of S in which X_i is mapped into σX_i. Thus S is a G-module. Indeed each S_ν is a finite-dimensional rational G-module. Thus, to sum up, G acts rationally on S by means of K-algebra automorphisms.

The first paragraph of the proof now shows that S^G is a Noetherian ring. But S^G is graded with grading

$$S^G = S_0^G \oplus S_1^G \oplus S_2^G \oplus \dots$$

and $S_0^G = K$. It therefore follows, from Lemma 13 Cor., that S^G is a finitely generated K-algebra.

Finally there is a surjective homomorphism $S \to R$ of K-algebras in which $X_i \mapsto u_i$. This is also a homomorphism of G-modules. Consequently, by Lemma 11, S^G gets mapped into R^G and indeed the induced mapping $S^G \to R^G$ is surjective. But S^G is a finitely generated K-algebra and therefore the same must hold for R^G. The proof is now complete.

5.11 Quotients with respect to linearly reductive groups

Let V be a non-empty affine set and G an affine group, both defined over K, and suppose that G acts morphically on the right of V. Now suppose that we have an affine set V_0 and a K-morphism

$$\pi : V \to V_0$$

such that $\pi(v\sigma) = \pi(v)$ for all $v \in V$ and $\sigma \in G$. We say that (V_0, π) is a quotient of V with respect to G or a quotient of V for the action of G if the following condition is satisfied. Given any K-morphism $\pi' : V \to V_0'$ of V into an affine set V_0' such that $\pi'(v\sigma) = \pi'(v)$ for all v and σ, there exists a unique K-morphism $\phi : V_0 \to V_0'$ such that $\phi \circ \pi = \pi'$.

Suppose that (V_0, π) and (V_0', π') are both of them quotients of V with respect to G and let $\phi_1 : V_0 \to V_0'$ and $\phi_2 : V_0' \to V_0$ be the K-morphisms such that $\phi_1 \circ \pi = \pi'$ and $\phi_2 \circ \pi' = \pi$. Then $\phi_2 \circ \phi_1$ and $\phi_1 \circ \phi_2$ are the identity mappings of V_0 and V_0' respectively. Thus ϕ_1 is a K-isomorphism and ϕ_2 is its inverse. Consequently if V has a quotient with respect to G, then the quotient is essentially unique.

The question of the existence of quotients is more difficult. Let $\sigma \in G$. There is a K-automorphism of V in which $v \mapsto v\sigma$ and corresponding to this we have an automorphism of the K-algebra K[V] in which $f \mapsto f^\sigma$, where $f^\sigma(v) = f(v\sigma)$. Put $\sigma f = f^\sigma$. Then, because G acts on the right of V, this turns K[V] into a left (G, K)-module. Suppose that $f \in K[V]$. Then, by Theorem 25 adapted to the case of right actions, the

K-space spanned by the elements $\{\sigma f\}_{\sigma \in G}$ has finite dimension. This K-space is also a rational G-module containing f as may be seen by adapting the proof of Lemma 4. Thus, to sum up, G *acts rationally on* K[V] *by means of* K-*algebra automorphism.*

The K-subalgebra $K[V]^G$, of K[V], consisting of the G-invariant coordinate functions on V is *rationally reduced.* This is because K[V] is rationally reduced and every rational maximal ideal of K[V] contracts to a rational maximal ideal of $K[V]^G$ (see Chapter 2 Theorem 1). Thus we obtain

Lemma 14. *Let the situation be as described above. Then* $K[V]^G$ *is an affine* K-*algebra if and only if it is finitely generated as a* K-*algebra.*

Suppose that $v \in V$ and let M_v denote the rational maximal ideal of K[V] corresponding to v. Then $f \in M_{v\sigma}$ if and only if $f^\sigma \in M_v$. Consequently when $f \in K[V]^G$ we have $f \in M_{v\sigma}$ if and only if $f \in M_v$. Accordingly

$$M_v \cap K[V]^G = M_{v\sigma} \cap K[V]^G \tag{5.11.1}$$

for all $v \in V$ and $\sigma \in G$.

Lemma 15. *Suppose that* $K[V]^G$ *is a finitely generated* K-*algebra. Then there exists a quotient of* V *with respect to* G. *Furthermore if* $\pi : V \to V_0$ *is such a quotient, then, in the corresponding homomorphism* $\pi^* : K[V_0] \to K[V]$ *of* K-*algebras,* $K[V_0]$ *is mapped isomorphically on to* $K[V]^G$.

Proof. By Lemma 14, $K[V]^G$ is an affine algebra. Consequently we can find an affine set V_0 such that $K[V_0]$ is isomorphic, as a K-algebra, to $K[V]^G$. We can therefore construct a homomorphism $\pi^* : K[V_0] \to K[V]$, of K-algebras, which maps $K[V_0]$ isomorphically on to $K[V]^G$ and this in turn will induce a morphism $\pi : V \to V_0$ of affine sets. Moreover, if $v \in V$ and $\sigma \in G$, then $\pi(v) = \pi(v\sigma)$ by virtue of (5.11.1).

Now let $\lambda : V \to U$ be a K-morphism of affine sets which is such that $\lambda(v\sigma) = \lambda(v)$ for all $v \in V$ and $\sigma \in G$. This determines a homomorphism $\lambda^* : K[U] \to K[V]$ of K-algebras. Let $h \in K[U]$ and put $f = \lambda^*(h)$. Then

$$f^\sigma(v) = f(v\sigma) = h(\lambda(v\sigma)) = h(\lambda(v)) = f(v)$$

and therefore $f^\sigma = f$ for all $\sigma \in G$. This shows that $\lambda^*(h) \in K[V]^G$ and thus we see that there is a unique K-algebra homomorphism $\omega^* : K[U] \to K[V_0]$ such that $\pi^* \circ \omega^* = \lambda^*$. But this is equivalent to saying that there is a unique K-morphism $\omega : V_0 \to U$ such that $\omega \circ \pi = \lambda$.

This establishes that $\pi : V \to V_0$ is a quotient of V with respect to G. Since quotients are essentially unique, the lemma follows.

The next corollary is more or less a restatement of the lemma.

Corollary 1. Let the affine group G act morphically on the right of the affine set V and let $\pi : V \to V_0$ be an almost surjective K-morphism of affine sets. If now $K[V_0]$, when considered as a subalgebra of $K[V]$, coincides with $K[V]^G$, then (V_0, π) is a quotient of V with respect to G.

Corollary 2. Suppose that the affine group G is linearly reductive Then V possesses a quotient with respect to G and, if $\pi : V \to V_0$ is such a quotient, the associated K-algebra homomorphism $\pi^* : K[V_0] \to K[V$ maps $K[V_0]$ isomorphically on to $K[V]^G$. If, in addition, K is algebrai ally closed, then π is surjective.

Proof. By Theorem 42, $K[V]^G$ is a finitely generated K-algebra. Consequently we need only establish the final assertion.

Suppose then that K is algebraically closed, and let N be a ration maximal ideal of $K[V]^G$. We have to show that N is the contraction of a rational maximal ideal of $K[V]$. But, because K is algebraically closed every maximal ideal of $K[V]$ is rational and therefore we need only show that $NK[V] \neq K[V]$. However this is clear because, by Lemma 12 Cor. 1, $NK[V] \cap K[V]^G = N$.

Until we come to the next lemma, we shall assume that our group C is linearly reductive and that the ground field K is algebraically closed. As before $\pi : V \to V_0$ denotes a quotient of V with respect to G.

First suppose that $\mathfrak{U}$ is an ideal of $K[V]$ and that W is its locus. Then $\pi^{*-1}(\mathfrak{U})$ is an ideal of $K[V_0]$. The locus of this will be X say. If now $g \in K[V_0]$ and g vanishes at all points of the closure $\overline{\pi(W)}$, of $\pi(W)$, then $\pi^*(g)$ vanishes on W and therefore $\pi^*(g^m) \in \mathfrak{U}$ for some positive integer m. The converse also holds. Thus g vanishes everywhere on $\overline{\pi(W)}$ if and only if some power of g belongs to $\pi^{*-1}(\mathfrak{U})$. Since K is algebraically closed, this means that g vanishes everywhere on $\overline{\pi(W)}$ if and only if it vanishes everywhere on X. Consequently $X = \overline{\pi(W)}$ that is to say $\overline{\pi(W)}$ is the locus, on V_0, of the ideal

$$\pi^{*-1}(\mathfrak{U}) = \pi^{*-1}(\mathfrak{U} \cap K[V]^G).$$

A subset W of V will be said to be G-invariant if $W\sigma = W$ for all $\sigma \in G$ or, equivalently, if W is a union of orbits. Suppose that W is a closed G-invariant subset of V and put $\mathfrak{U} = I_V(W)$. If now $f \in \mathfrak{U}$, $\sigma \in G$ and $w \in W$, then

$$f^\sigma(w) = f(w\sigma) = 0$$

and therefore $f^\sigma \in \mathfrak{U}$. This shows that the ideal $\mathfrak{U}$ is a G-submodule of $K[V]$.

Lemma 16. Let G be linearly reductive, K algebraically closed, and $\pi : V \to V_0$ a quotient of V with respect to G. If now $\{W_j\}_{j \in J}$ is a family of closed G-invariant subsets of V, then

$$\overline{\pi(\bigcap_{j \in J} W_j)} = \bigcap_{j \in J} \overline{\pi(W_j)}\ .$$

Also if X is a closed G-invariant subset of V, then $\pi(X)$ is closed in V_0.

Proof. Put $\mathfrak{U}_j = I_V(W_j)$ and

$$W = \bigcap_{j \in J} W_j\ .$$

Then W_j is the locus of $\mathfrak{U}_j$, $\overline{\pi(W_j)}$ is the locus of $\pi^{*-1}(\mathfrak{U}_j \cap K[V]^G)$, W is the locus of $\Sigma\mathfrak{U}_j$, and $\overline{\pi(W)}$ is the locus of $\pi^{*-1}((\Sigma\mathfrak{U}_j) \cap K[V]^G)$. However, $\mathfrak{U}_j$ is a G-submodule of $K[V]$ and therefore, by Lemma 12 Cor. 2,

$$(\sum_{j\in J} \mathfrak{U}_j) \cap K[V]^G = \sum_{j\in J} (\mathfrak{U}_j \cap K[V]^G).$$

We now see that $\overline{\pi(W)}$ is the locus of

$$\sum_{j\in J} \pi^{*-1}(\mathfrak{U}_j \cap K[V]^G)$$

and this shows that

$$\overline{\pi(W)} = \bigcap_{j\in J} \overline{\pi(W_j)}.$$

Now let X be a closed G-invariant subset of V, and let $v_0 \in \overline{\pi(X)}$. Put $X' = \pi^{-1}(\{v_0\})$. Then X' is also a closed G-invariant subset of V and therefore

$$\overline{\pi(X \cap X')} = \overline{\pi(X)} \cap \{v_0\} = \{v_0\}$$

by what has just been proved. It follows that $X \cap X'$ is not empty and therefore $v_0 \in \pi(X)$. Consequently $\pi(X)$ is closed in V_0.

We now introduce a concept which is stronger than that of a quotient. Suppose that the affine group G acts morphically on the right of the affine set V, and let $\pi : V \to V_0$ be a K-morphism of affine sets.

Definition. We say that (V_0, π) is a 'strict quotient' of V with respect to G provided that the following three conditions are satisfied:

(1) π is surjective and for each $y \in V_0$, $\pi^{-1}(\{y\})$ is an orbit of G in V;

(2) as a subalgebra of $K[V]$, $K[V_0]$ coincides with $K[V]^G$;

(3) $\pi : V \to V_0$ is an open mapping, that is to say open subsets of V are mapped on to open subsets of V_0.

Note that, by Lemma 15 Cor. 1, a strict quotient of V with respect to G is a quotient in the original sense.

Our principal result on quotients with respect to linearly reductive groups is due to D. Mumford and may be stated as follows.

Theorem 43. Suppose that the ground field K is algebraically closed, that G is a linearly reductive affine group, and that G acts morphically on the right of V. Suppose further that all the orbits of G

in V are closed. Then V possesses a strict quotient with respect to G.

Proof. By Lemma 15 Cor. 2, there is a quotient $\pi : V \to V_0$, of V with respect to G, where π is surjective and $K[V_0] = K[V]^G$.

Let $y \in V_0$. Then $\pi^{-1}(\{y\})$ is non-empty and it is a union of orbits. Let W, W' be orbits in this union. Then they are closed (by hypothesis) and they are G-invariant.

We claim that $W = W'$. For assume the contrary. Then $W \cap W'$ is empty and therefore

$$I_V(W) + I_V(W') = K[V]$$

because K is algebraically closed. Accordingly, by Lemma 12 Cor. 2,

$$(I_V(W) \cap K[V]^G) + (I_V(W') \cap K[V]^G) = K[V]^G$$

and therefore $f + g = 1$ for a suitable f in $I_V(W) \cap K[V]^G$ and g in $I_V(W') \cap K[V]^G$. Note that f vanishes everywhere on W and takes the value 1 at all points of W'. Furthermore $f = \pi^*(h)$ for some h in $K[V_0]$. But now if $x \in W$ and $x' \in W'$, then

$$f(x) = h(y) = f(x')$$

and thus we have a contradiction. This establishes our claim and shows that, for each $y \in V_0$, $\pi^{-1}(\{y\})$ is an orbit of G in V. It remains only for us to show that $\pi : V \to V_0$ is an open mapping.

To this end let T be an open subset of V and put $S = \pi(T)$. In view of what has just been proved we have

$$\pi^{-1}(S) = \bigcup_{\sigma \in G} T\sigma$$

and this is open in V. It follows that $\pi^{-1}(V_0 \backslash S)$ is a closed G-invariant subset of V and therefore, by Lemma 16, $\pi(\pi^{-1}(V_0 \backslash S)) = V_0 \backslash S$ is closed in V_0. But this means that S is open in V_0 and with this the proof is complete.

5.12 Quotients with respect to finite groups

The discussion of quotients set out in section (5.11) deals primarily with linearly reductive groups, and therefore it does not cover the case of *finite* groups. These are easier to handle and will be given an *ad hoc* treatment in this section.

Let R be a non-trivial, unitary, associative and commutative K-algebra. (For the moment K is an arbitrary field.) Let G be a *finite* group which acts on R by means of K-algebra automorphisms so we have, in effect, a homomorphism of G into the group of K-algebra automorphisms of R. Those elements of R that are left fixed by the elements of G form the K-subalgebra R^G. If $f \in R$, then the effect of operating on f with the element σ of G will be denoted by f^σ. Note that $(f^\sigma)^\tau = f^{\tau\sigma}$.

Now suppose that X is an indeterminate. Each σ in G induces, in an obvious manner, an automorphism of the polynomial ring R[X]. Also, for $f \in R$, the polynomial

$$\prod_{\sigma \in G} (X - f^\sigma)$$

belongs to R[X] and is invariant under the automorphisms induced by the members of G. It follows that its coefficients, say $a_1, a_2, \ldots, a_n$, all belong to R^G. Since the polynomial is monic and has f as a root, it follows that f is integral over $K[a_1, a_2, \ldots, a_n]$ and therefore it is integral over the larger algebra R^G. In particular we see that R *is an integral extension of* R^G.

Theorem 44. *Let G be a finite group and let the situation be as described above. If now R is a finitely generated K-algebra, then R^G is also a finitely generated K-algebra. Furthermore under these conditions R is not only an integral extension of R^G but it is in fact a finitely generated R^G-module. Finally every maximal ideal of R^G is the contraction of a maximal ideal of R.*

Proof. Let $R = K[f_1, f_2, \ldots, f_s]$ and denote by $a_{i1}, a_{i2}, \ldots, a_{in}$ the coefficients of the polynomial

$$\prod_{\sigma \in G} (X - f_i^{\sigma}).$$

Then $a_{ij} \in R^G$. Now put

$$R_0 = K[a_{11}, a_{12}, \ldots, a_{sn}].$$

Evidently R_0 is a Noetherian K-subalgebra of R^G and each f_i is integral over R_0. It follows† that $R = R_0[f_1, f_2, \ldots, f_s]$ is a finitely generated R_0-module. However this implies that R^G is a finitely generated K-algebra. All the assertions of the theorem are now clear except possibly the one concerning maximal ideals. This is true simply because R is an integral extension of R^G.

We now turn our attention to affine sets. Suppose therefore that V is a non-empty affine set defined over K and that G is a group that acts on the _right_ of V. We recall that if G is _finite_, then it is automatically an affine group.

Lemma 17. _Suppose that G is a finite group. Then in order that the action of G on V be K-morphic it is necessary and sufficient that, for each $\sigma \in G$, the mapping $V \to V$ in which $v \mapsto v\sigma$ be a K-morphism._

Proof. The condition is clearly necessary. Now assume that it is satisfied. For each $\tau \in G$ define $h_\tau : G \to K$ by $h_\tau(\sigma) = \delta_{\tau\sigma}$, where δ denotes Kronecker's function. Then $h_\tau \in K[G]$ because G is finite.

Next assume that $f \in K[V]$ and define $f^\tau : V \to K$ by $f^\tau(v) = f(v\tau)$. Our assumptions ensure that $f^\tau \in K[V]$ and therefore

$$\sum_{\tau \in G} f^\tau \vee h_\tau \in K[V \times G].$$

Moreover the value of this particular function at (v, σ) is $f(v\sigma)$. Consequently the mapping $V \times G \to V$ which is the action of G on V is a K-morphism and the lemma follows.

We shall now assume that V is a non-empty affine set defined over an _algebraically closed_ field K and that G is a _finite group_ which acts morphically on the right of V. These assumptions will remain in force until the proof of Lemma 18 has been completed.

---

† For an account of the theory of integral extensions see, for example, [(9) pp. 86-93].

As we saw in section (5.11), these conditions ensure that G acts rationally on $K[V]$ by means of K-algebra automorphisms. Next Theorem 44 shows that $K[V]^G$ is a finitely generated K-algebra and therefore, by Lemma 15, there exists a quotient $\pi : V \to V_0$ of V with respect to G. The same lemma also shows that the associated K-algebra homomorphism $\pi^* : K[V_0] \to K[V]$ maps $K[V_0]$ isomorphically on to $K[V]^G$. Finally $\pi : V \to V_0$ is a surjection. This is a consequence of the final assertion of Theorem 44 and our assumption that K is algebraically closed.

Lemma 18. *Let the situation be as described above. Then for every closed subset U, of V, $\pi(U)$ is a closed subset of V_0, that is π is a closed mapping.*

Proof. Let $\mathfrak{U} = I_V(U)$, $\mathfrak{B} = \mathfrak{U} \cap K[V]^G$, and let X be the locus of $\pi^{*-1}(\mathfrak{B})$ in V_0. Then $K[V]^G/\mathfrak{B}$ is a subring of $K[V]/\mathfrak{U}$ and the latter is an integral extension of the former. It follows that every maximal ideal of $K[V]^G/\mathfrak{B}$ is the contraction of a maximal ideal of $K[V]/\mathfrak{U}$ and therefore every maximal ideal of $K[V]^G$ containing $\mathfrak{B}$ is the contraction of a maximal ideal of $K[V]$ containing $\mathfrak{U}$. On the other hand, every maximal ideal of $K[V]$ that contains $\mathfrak{U}$ contracts, in $K[V]^G$, to a maximal ideal containing $\mathfrak{B}$. If all this is translated into geometrical terms it says simply that $\pi(U) = X$. In particular $\pi(U)$ is closed in V_0.

Theorem 45. *Let V be a non-empty affine set defined over an algebraically closed field K and let G be a finite group which acts K-morphically on the right of V. Then V possesses a strict quotient with respect to G.*

Proof. We know that there exists a quotient $\pi : V \to V_0$, of V with respect to G, where π is a surjection and $\pi^* : K[V_0] \to K[V]$ maps $K[V_0]$ isomorphically on to $K[V]^G$. Furthermore, by Lemma 18, π takes closed subsets of V on to closed subsets of V_0.

Let $y \in V_0$. Then $\pi^{-1}(\{y\})$ is non-empty and it is a union of orbits. Let W and W' be orbits contained in $\pi^{-1}(\{y\})$.

We claim that $W = W'$. For suppose that $W \neq W'$. Then $W \cap W'$

is empty and each of W and W' is a finite set and therefore closed in V. It follows, because K is algebraically closed, that

$$I_V(W) + I_V(W') = K[V]$$

and therefore we can find $h \in I_V(W)$ and $h' \in I_V(W')$ such that $h + h' = 1$. Of course h takes the value 0 at all points of W and the value 1 at all points of W'.

Let $x \in W$, $x' \in W'$ and put

$$h_0 = \prod_{\sigma \in G} h^\sigma.$$

Then $h_0 \in K[V]^G$ and therefore $h_0 = \pi^*(f)$ for some $f \in K[V_0]$. Also $h_0(x) = 0$ and

$$h_0(x') = \prod_{\sigma \in G} h(x'\sigma) = 1$$

because $x'\sigma \in W'$ for all $\sigma \in G$. However $h_0(x) = f(\pi(x)) = f(y) = f(\pi(x')) = h_0(x')$ and this is the desired contradiction. Thus our claim is established and it follows that $\pi^{-1}(\{y\})$ is an orbit.

It only remains for us to show that $\pi : V \to V_0$ is an open mapping. However this is no problem because the argument used to complete the proof of Theorem 43 works again here.

An example. Let V be a non-empty affine set defined over an algebraically closed field K and let n $(n \geq 1)$ be an integer. Put

$$W = V \times V \times \ldots \times V \quad (n \text{ factors})$$

and let G be the permutation group of $\{1, 2, \ldots, n\}$, i.e. G is the symmetric group of degree n.

For $w = (v_1, v_2, \ldots, v_n)$ in W and σ in G define $w\sigma$ by

$$w\sigma = (v_{\sigma(1)}, v_{\sigma(2)}, \ldots, v_{\sigma(n)}).$$

This leads to a <u>right</u> action of G on W and since, for fixed $\sigma \in G$, $w \mapsto w\sigma$ is a K-automorphism of the affine set W, Lemma 17 shows that G acts K-morphically on W. Accordingly, by Theorem 45, W possesses a strict quotient $\pi : W \to W_0$ with respect to G.

Let us examine this quotient. The points of W_0 are matched with the orbits of G in V. Also two points $(v_1, v_2, \ldots, v_n)$ and $(v'_1, v'_2, \ldots, v'_n)$ of W belong to the same orbit if and only if $(v'_1, v'_2, \ldots, v'_n)$ is a permutation of $(v_1, v_2, \ldots, v_n)$. Thus the orbits of G are the unordered n-tuples of points of V. Accordingly these unordered n-tuples may be regarded, in a natural way, as the points of an affine set W_0. When this is done the associated surjection $\pi : W \to W$ is a closed morphism, and the coordinate ring of W_0 consists of all $f \in K[W]$ with the property that

$$f(v_1, v_2, \ldots, v_n) = f(v'_1, v'_2, \ldots, v'_n)$$

whenever $v'_1, v'_2, \ldots, v'_n$ is a permutation of $v_1, v_2, \ldots, v_n$.

5.13 Quotient groups of affine groups

Let G be an affine group and H a closed, normal subgroup of G. We shall now enquire whether the quotient group G/H has a natural structure as an affine group and, if so, what is the coordinate ring of G/H and what are the properties of the natural homomorphism $G \to G/H$? Answers to these questions will be found in Theorem 48 and, roughly speaking, all one might reasonably hope for is realised, at least in the case where the ground field is algebraically closed. However we shall only arrive at these conclusions after some comparatively lengthy considerations and we shall need to draw rather more heavily on the theory of commutative Noetherian rings than we have done so far.

We begin by supplementing our previous results on affine sets.

Lemma 19. *Let V be an affine set defined over an algebraically closed field K and let $V_1, V_2, \ldots, V_s$ be disjoint closed subsets of V whose union is V itself. Then K[V] can be regarded as the direct product*

$$K[V_1] \times K[V_2] \times \ldots \times K[V_s]$$

of K-algebras with the projection $K[V] \to K[V_i]$ corresponding to the inclusion mapping of V_i into V.

Proof. Put $W_i = V_1 \cup \ldots \cup V_{i-1} \cup V_{i+1} \cup \ldots \cup V_s$. Then no maximal ideal of $K[V]$ contains both $I_V(V_i)$ and $I_V(W_i)$ and therefore $\eta_i + \varepsilon_i = 1$ for suitable functions $\eta_i \in I_V(V_i)$ and $\varepsilon_i \in I_V(W_i)$. Note that ε_i takes the value 1 everywhere on V_i and the value 0 at all other points of V; that $\varepsilon_1, \varepsilon_2, \ldots, \varepsilon_s$ are orthogonal idempotents of $K[V]$; and that $\varepsilon_1 + \varepsilon_2 + \ldots + \varepsilon_s = 1$. Put $A_i = K[V]\varepsilon_i$. Then A_i is a K-algebra and $K[V]$ is the direct product of $A_1, A_2, \ldots, A_s$.

We must now identify A_i. The K-algebra homomorphism $K[V] \to A_i$ in which $f \mapsto f\varepsilon_i$ is surjective and its kernel is $I_V(V_i)$. Hence we have $A_i = K[V_i]$ by identifying $f\varepsilon_i$ with the restriction of f to V_i, and then the projection homomorphism of $K[V]$ on to $A_i = K[V_i]$ corresponds to the inclusion $V_i \subseteq V$.

The lemma also has a converse.† For suppose that affine sets $V_1, V_2, \ldots, V_s$ are given and let us form the direct product

$$K[V_1] \times K[V_2] \times \ldots \times K[V_s].$$

This is an affine K-algebra. Also its rational maximal ideals correspond to the rational maximal ideals of the various factors and so they are in a natural one-one correspondence with the points of the disjoint union, V say, of $V_1, V_2, \ldots, V_s$. Thus we may regard V as an affine set with

$$K[V] = K[V_1] \times K[V_2] \times \ldots \times K[V_s].$$

Let $\mathfrak{U}_i$ be the kernel of the projection homomorphism $K[V] \to K[V_i]$. The rational maximal ideals of $K[V]$ that contain $\mathfrak{U}_i$ correspond to the points of V_i. Thus $V_i = C_V(\mathfrak{U}_i)$ and therefore V_i is closed when regarded as a subset of V. Also, because $K[V_i]$ is rationally reduced, $\mathrm{Rad}_V(\mathfrak{U}_i) = \mathfrak{U}_i$ and therefore $I_V(V_i) = \mathfrak{U}_i$. It follows from this that the affine structure of V_i is not changed by regarding it as a closed subset of V, and that the projection of $K[V]$ on to $K[V_i]$ corresponds to the inclusion mapping of V_i into V.

Now let $V'_1, V'_2, \ldots, V'_s$ be additional affine sets defined over K.

---

† This part of the argument does not require K to be algebraically closed.

Their disjoint union, V' say, may be regarded as an affine set with

$$K[V'] = K[V'_1] \times K[V'_2] \times \ldots \times K[V'_s].$$

Assume that we are given K-morphisms $\pi_i : V_i \to V'_i$, where $i = 1, 2, \ldots, s$. The corresponding K-algebra homomorphisms $\pi_i^* : K[V'_i] \to K[V_i]$ induce a K-algebra homomorphism

$$\pi^* : K[V'] \to K[V]$$

and this corresponds to a K-morphism $\pi : V \to V'$. Observe that, for each i, π is an extension of $\pi_i : V_i \to V'_i$.

Our next result extends Theorem 38 of Chapter 3.

Theorem 46. _Let_ V _and_ W _be irreducible affine sets defined over an algebraically closed field_ K. _If now_ $\phi : V \to W$ _is both an almost surjective_ K-_morphism and an injection, then_ K(V) _is a purely inseparable, algebraic extension of_ K(W).

Proof. The hypotheses ensure that K(W) is a subfield of the field K(V) and we know, from Chapter 3 Theorem 38, that Dim V = Dim W. Since K[V] is a finitely generated K-algebra, it follows that K(V) is an algebraic extension of K(W) of finite degree.

Let Ω be the separable closure of K(W) in K(V). _We claim that_ $\Omega = (K(W))(f)$ _for some_ $f \in K[V]$. In establishing this claim we shall suppose that the characteristic of K is the prime p. (If K has characteristic zero, then the proof is similar but without certain complications.)

Certainly $\Omega = (K(W))(h)$ for some $h \in K(V)$ and

$$\Omega = (K(W))(h^{p^n})$$

for every positive integer n. Now write h as the quotient of two elements of K[V]. Then, provided n is large enough, h^{p^n} is the quotient of two elements both of which belong to $K[V] \cap \Omega$. Accordingly $\Omega = (K(W))(f_1, f_2)$ where f_1, f_2 are in K[V]. But there are only finitely many fields between K(W) and Ω, and the field K is infinite. It follows that

$$\Omega = (K(W))(f_1 + cf_2)$$

for a suitable $c \in K$. Since $f_1 + cf_2$ is in $K[V]$, this establishes our claim.

By multiplying f by a suitable non-zero element of $K[W]$ we can arrange for it to have an additional property. Thus if

$$F(X) = X^n + \omega_1 X^{n-1} + \omega_2 X^{n-2} + \ldots + \omega_n$$

is the irreducible polynomial for f over $K[W]$, we can secure that the ω_i are in $K[W]$.

The polynomial $F(X)$ is irreducible and separable over $K(W)$ and therefore $F(X)$ and its formal derivative $F'(X)$ satisfy a relation

$$F(X)G(X) + F'(X)H(X) = q,$$

where $G(X)$, $H(X)$ are in $(K[W])[X]$ and $q \in K[W]$, $q \neq 0$.

Consider the K-algebras

$$K[W] \subseteq (K[W])[f] \subseteq K[V].$$

By Chapter 3 Lemma 8, there exists $g \in (K[W])[f]$, $g \neq 0$ such that every maximal ideal of $(K[W])[f]$ not containing g is the contraction of a maximal ideal of $K[V]$. Also our choice of f ensures that $(K[W])[f]$ is an integral extension of $K[W]$ and therefore g satisfies a relation

$$g^m + \gamma_1 g^{m-1} + \gamma_2 g^{m-2} + \ldots + \gamma_m = 0,$$

where $\gamma_i \in K[W]$. Since $g \neq 0$ we may suppose that $\gamma_m \neq 0$ and then it follows that every maximal ideal of $K[W]$ not containing γ_m is the contraction of a maximal ideal of $(K[W])[f]$ not containing g, and this in turn will be the contraction of a maximal ideal of $K[V]$.

Choose $w \in W$ so that $q(w) \neq 0$ and $\gamma_m(w) \neq 0$. This determines a maximal ideal M of $K[W]$. But $\gamma_m \notin M$ so M is the contraction of a maximal ideal of $K[V]$. However $\phi : V \to W$ is an injection. Consequently there is only one maximal ideal of $K[V]$ that contracts to M. It follows that _there is exactly one maximal ideal of_ $(K[W])[f]$ _that contracts to_ M.

The homomorphism $K[W] \to K$ of K-algebras in which $p \mapsto p(w)$ induces a homomorphism

$$(K[W])[X] \to K[X]$$

of the polynomial rings. For $Q(X) \in (K[W])[X]$ let us use $\overline{Q}(X)$ to denote its image. Then

$$\overline{F}(X)\overline{G}(X) + \overline{F}'(X)\overline{H}(X) = q(w)$$

and therefore, since $q(w) \neq 0$, $\overline{F}(X)$ has n distinct roots, where n is the degree of $F(X)$. Let α be any one of these roots. Then the homomorphism

$$(K[W])[X] \to K$$

in which $Q(X) \mapsto \overline{Q}(\alpha)$ induces a homomorphism $(K[W])[f] \to K$ in which $f \mapsto \alpha$. Thus there are n different maximal ideals of $(K[W])[f]$ that contract to M and therefore $n = 1$. However this means that $\Omega = K(W)$ and so the proof is complete.

Before we leave this preliminary section on affine sets, it is convenient to put on record an elementary result concerning finitely generated field extensions.

Lemma 20. Let E be an extension field of a field F and let L be a field between F and E. If now E is a finitely generated field extension of F, then so too is L.

Proof. Let $\xi_1, \xi_2, \ldots, \xi_s$ be a transcendence base for L over F and let this be extended to a transcendence base $\xi_1, \ldots, \xi_s, \eta_1, \ldots, \eta_t$ for E over F. Note that $\eta_1, \eta_2, \ldots, \eta_t$ will be algebraically independent over L.

Let $\lambda_1, \lambda_2, \ldots, \lambda_q$ belong to L and be linearly independent over $F(\xi_1, \xi_2, \ldots, \xi_s)$. These elements will remain linearly independent over $F(\xi_1, \ldots, \xi_s, \eta_1, \ldots, \eta_t)$. Hence, if m is the degree (necessarily finite) of E over $F(\xi_1, \ldots, \xi_s, \eta_1, \ldots, \eta_t)$, we must have $q \leq m$. Thus the degree of L over $F(\xi_1, \xi_2, \ldots, \xi_s)$ is finite and therefore L

is a finitely generated field extension of F.

We are now ready to begin the investigation of quotient groups of affine groups. To avoid excessive repetition we shall make the following assumptions which are to remain in force until we come to the statement of Lemma 25:

(1) *the field* K *is algebraically closed;*

(2) G *is a connected affine group defined over* K;

(3) H *is a closed, normal subgroup of* G;

(4) *there is given a connected affine group* Γ *and a surjective* K-*homomorphism* $\phi : G \to \Gamma$ *such that* $\operatorname{Ker} \phi = H$.

Note that if (1), (2) and (3) are satisfied, then we can always find ϕ and Γ so that (4) holds as well. For, by Lemma 10, there exists a finite-dimensional rational left G-module such that if $\rho : G \to GL(V)$ is the corresponding rational representation, then the kernel of ρ is H. Put $\Gamma = \rho(G)$. By Theorem 22, Γ is a closed subgroup of GL(V) and therefore (4) is satisfied if ϕ is taken to be the induced K-homomorphism $G \to \Gamma$. Of course Γ will be connected because it is a continuous image of G.

The mapping $G \times H \to G$ in which $(\sigma, \tau) \mapsto \sigma\tau$ provides a K-morphic action by H on the right of G. Consequently H acts rationally on K[G] by means of K-algebra automorphisms. As in similar situations we use $K[G]^H$ to denote the subalgebra formed by the functions that H leaves invariant. Since ϕ is a surjective K-homomorphism with kernel H, we have

$$K[\Gamma] \subseteq K[G]^H \subseteq K[G].$$

Moreover K[G] is an integral domain because G is connected.

Lemma 21. *The quotient field of* $K[G]^H$ *is a purely inseparable, algebraic extension of* $K(\Gamma)$.

Proof. By Lemma 20, we can choose $f_1, f_2, \ldots, f_s$ in $K[G]^H$ so that

$$K[\Gamma] \subseteq K[f_1, f_2, \ldots, f_s] \subseteq K[G]^H \subseteq K[G] \tag{5.13.1}$$

and $K[f_1, f_2, \ldots, f_s]$ has the same quotient field as $K[G]^H$. Evidently $K[f_1, f_2, \ldots, f_s]$ is an affine K-algebra. Consequently there exists an irreducible affine set V with $K[V] = K[f_1, f_2, \ldots, f_s]$ and then the inclusions in (5.13.1) will give rise to almost surjective K-morphisms $\theta : G \to V$ and $\psi : V \to \Gamma$ such that $\psi \circ \theta = \phi$. By Chapter 3 Theorem 33 $\theta(G)$ contains a non-empty open subset U, of V, and now, by making U smaller if necessary, we may suppose that U consists of the points of V where some member of $K[V]$ does not vanish. This ensures (see Chapter 2 Theorem 20) that U has a natural structure as an affine set with $K[V] \subseteq K[U]$ and $K(V) = K(U)$.

Next, again by Chapter 3 Theorem 33, $\psi(U)$ contains a non-empty open subset of Γ. Accordingly the mapping $U \to \Gamma$ induced by ψ is an almost surjective K-morphism.

Suppose that $f \in K[V] \subseteq K[G]^H$. Then $f(\sigma\tau) = f(\sigma)$ for all $\sigma \in G$ and $\tau \in H$, and from this it follows that $\theta : G \to V$ is constant on the cosets of H in G. Suppose that $u_1, u_2 \in U$, say $u_i = \theta(\sigma_i)$, and that $\psi(u_1) = \psi(u_2)$. Then $\phi(\sigma_1) = \phi(\sigma_2)$. Thus σ_1 and σ_2 belong to the same coset of H and therefore $u_1 = u_2$. Thus our K-morphism $U \to \Gamma$ is an injection as well as being almost surjective. Consequently, by Theorem 46, $K(U) = K(V)$ is a purely inseparable, algebraic extension of $K(\Gamma)$. The lemma follows.

Lemma 22. <u>With the above assumptions</u>

$$K(\Gamma) \cap K[G]^H = K[\Gamma].$$

Proof. Suppose that $f \in K(\Gamma) \cap K[G]^H$ and let $y \in \Gamma$. By (3.3.5) we can regard the local ring $Q_{\Gamma, y}$, of Γ at y, as lying between $K[\Gamma]$ and $K(\Gamma)$. On this understanding <u>we claim that</u> f <u>belongs to</u> $Q_{\Gamma, y}$. For assume that $f \notin Q_{\Gamma, y}$. By Theorem 13, y is a <u>simple</u> point of Γ. Certainly $Q_{\Gamma, y}$ is a Noetherian ring with a single maximal ideal and now we see, by Chapter 4 Theorem 23, that it is what is called a <u>regular</u> local ring. Now it is known[†] that every regular local ring is a unique

† See [(13) Vol. 2, Appendix 7, p. 404] or [(12) Theorem 5, p. 22].

factorization domain. This fact is used in the argument that follows.

Write $f = f'/f''$, where f', f'' are in $Q_{\Gamma,y}$ and have no common irreducible factor, and then, by multiplying numerator and denominator by a suitable unit of $Q_{\Gamma,y}$, arrange that $f', f'' \in K[\Gamma]$. Since $f \notin Q_{\Gamma,y}$, f'' has at least one irreducible factor. If ω is such a factor, then $\omega Q_{\Gamma,y} = PQ_{\Gamma,y}$, where P is a prime ideal of $K[\Gamma]$, $f'' \in P$ whereas $f' \notin P$. This shows that f' is not in the radical of the ideal $f''K[\Gamma]$ and therefore there exists $z \in \Gamma$ such that $f''(z) = 0$ and $f'(z) \neq 0$. Choose $\sigma \in G$ so that $\phi(\sigma) = z$.

Now $ff'' = f'$ and we can regard all of f, f', f'' as belonging to $K[G]$ in which case $f(\sigma)f''(\sigma) = f'(\sigma)$. However the value of f' at σ is the same as its value at z because of the way we embed $K[\Gamma]$ in $K[G]$. Consequently $f'(\sigma) \neq 0$. Similar considerations show that $f''(\sigma) = 0$. The relation $f(\sigma)f''(\sigma) = f'(\sigma)$ now yields a contradiction and with it our claim that $f \in Q_{\Gamma,y}$ is established. Accordingly we can find $\theta_y \in K[\Gamma]$ such that $\theta_y f \in K[\Gamma]$ and $\theta_y(y) \neq 0$.

Let us choose such a θ_y for each y in Γ. The ideal they generate in $K[\Gamma]$ is $K[\Gamma]$ itself because there is no single point where they all vanish. Since $\theta_y f \in K[\Gamma]$ for every y it follows that $f \in K[\Gamma]$.

It has now been proved that $K[\Gamma] \supseteq K(\Gamma) \cap K[G]^H$ and, as the reverse inclusion is trivial, this completes the proof.

Lemma 23. _The algebra_ $K[G]^H$ _is an integral extension of_ $K[\Gamma]$ _and each prime ideal of_ $K[\Gamma]$ _is the contraction of exactly one prime ideal of_ $K[G]^H$. _Furthermore_ $K[G]^H$ _is an affine_ K-_algebra._

Proof. For the moment we shall leave the final assertion on one side.

First suppose that K has characteristic zero. In this case Lemma 21 shows that $K(\Gamma)$ contains $K[G]^H$ and therefore $K[\Gamma] = K[G]^H$. Thus our assertions are trivial. Now assume that the characteristic of K is the prime p and let $f \in K[G]^H$. By Lemma 21, f^{p^n} is in $K(\Gamma) \cap K[G]^H = K[\Gamma]$ provided that n is large enough. Hence in this case too the first two assertions are clear. Note that we have established more about the relation between $K[G]^H$ and $K[\Gamma]$ than is actually stated in the lemma.

Indeed the extra information will be needed later.

Let us now turn our attention to the final assertion. By Lemmas 20 and 21, the quotient field of $K[G]^H$ is an algebraic extension of $K(\Gamma)$ of finite degree. Consequently, because $K[\Gamma]$ is a finitely generated K-algebra, the integral closure of $K[\Gamma]$ in this quotient field is a finitely generated $K[\Gamma]$-module† which has $K[G]^H$ as a $K[\Gamma]$-submodule. It follows, because $K[\Gamma]$ is Noetherian, that $K[G]^H$ is also a finitely generated $K[\Gamma]$-module and therefore it must be a finitely generated K-algebra. Finally $K[G]^H$ is rationally reduced because it is a K-subalgeb of $K[G]$. With this the proof is complete.

We are now ready to take a major step forward. Let us regard G as an affine set with H acting morphically on its right.

Lemma 24. <u>There exists a quotient</u> $\pi : G \to S$ <u>for the right action of</u> H <u>on</u> G. <u>Such a quotient is surjective and its fibres are precisely the cosets of</u> H <u>in</u> G. <u>Furthermore if we regard</u> $K[S]$ <u>as a subalgebra of</u> $K[G]$, <u>then</u> $K[S] = K[G]^H$.

Proof. By Lemma 15, there exists a quotient $\pi : G \to S$ of G with respect to H; moreover such a quotient will be almost surjective and it will lead to $K[S]$ being identified with $K[G]^H$. Next, since

$$K[\Gamma] \subseteq K[G]^H = K[S],$$

there exists a K-morphism $\lambda : S \to \Gamma$ which satisfies $\lambda \circ \pi = \phi$. Howeve Lemma 23 shows that λ is a bijection. The remaining assertions concerning π are now consequences of the corresponding properties of ϕ.

Let $\pi : G \to S$ be as in Lemma 24. Then there is a unique way of putting a group structure on S so that π becomes a surjective homomorphism of abstract groups whose kernel is H. What is not immediatel clear is that this will turn S into an <u>affine</u> group. The next lemma show that this is indeed the case. It should be noted that, in the lemma, we drop the requirement that G be <u>connected</u> because the more general version will be needed later. (We also do not require K to be algebraic

† See, for example, [(13) Vol. 1, Theorem 9, p. 267].

closed but this is less significant.) As usual our affine groups and sets are assumed to be defined over K.

Lemma 25. _Let G be a (not necessarily connected) affine group, H a closed normal subgroup of G, and_ $\pi : G \to S$ _a surjective K-morphism of affine sets whose fibres are precisely the cosets of H in G. If now K[S], when considered as a subalgebra of K[G], is just_ $K[G]^H$, _then there is a unique way to endow S with a group structure so that S becomes an affine group and_ π _a group homomorphism._

Remarks. Of course if S is given the structure referred to, then π becomes a surjective K-homomorphism of affine groups whose kernel is H. For G connected, S represents an improvement on Γ to the extent that it has the natural coordinate ring.

Proof. We turn S into a group in such a way that π is a homomorphism of groups and verify that, for S, multiplication and inversion are K-morphisms.

Let $\omega \in K[S]$. Since $\omega \circ \pi$ belongs to K[G], the mapping $h : G \times G \to K$ given by $h(\sigma_1, \sigma_2) = \omega(\pi(\sigma_1 \sigma_2))$ belongs to $K[G \times G]$. Consequently h can be represented in the form

$$h = \sum_{i=1}^{m} f_i \vee g_i, \tag{5.13.2}$$

where $f_i, g_i \in K[G]$. _We claim that we can arrange that all the_ f_i _are in_ $K[G]^H$. To see this we first make $g_1, g_2, \ldots, g_m$ linearly independent over K. Then, for $\sigma_1, \sigma_2 \in G$ and $\tau \in H$,

$$h(\sigma_1 \tau, \sigma_2) = \omega(\pi(\sigma_1 \tau \sigma_2)) = \omega(\pi(\sigma_1 \sigma_2)) = h(\sigma_1, \sigma_2)$$

whence

$$\sum_{i=1}^{m} f_i(\sigma_1 \tau) g_i(\sigma_2) = \sum_{i=1}^{m} f_i(\sigma_1) g_i(\sigma_2)$$

and therefore

$$\sum_{i=1}^{m} f_i(\sigma_1 \tau) g_i = \sum_{i=1}^{m} f_i(\sigma_1) g_i.$$

It follows that $f_i(\sigma_1 \tau) = f_i(\sigma_1)$ and hence that $f_i \in K[G]^H$. This establishes our claim.

Let us make a fresh start. Suppose that (5.13.2) holds but now with all the f_i in $K[G]^H$. Without destroying this property we can also suppose that $f_1, f_2, \ldots, f_m$ are linearly independent over K, and then an argument similar to that just used shows that the g_i are in $K[G]^H$ as well. Thus for each i we can find $u_i, v_i \in K[S]$ so that $u_i \circ \pi = f_i$ and $v_i \circ \pi = g_i$. By construction the value of

$$\sum_{i=1}^{m} u_i \vee v_i$$

at $(\pi(\sigma_1), \pi(\sigma_2))$ is $\omega(\pi(\sigma_1\sigma_2)) = \omega(\pi(\sigma_1)\pi(\sigma_2))$ and so the mapping $S \times S \to K$ in which

$$(\pi(\sigma_1), \pi(\sigma_2)) \mapsto \omega(\pi(\sigma_1)\pi(\sigma_2))$$

belongs to $K[S \times S]$. However, because ω is an arbitrary element of $K[S]$, this is equivalent to saying that the multiplication mapping $S \times S \to S$ is a K-morphism.

Next the mapping $p : G \to K$ given by $p(\sigma) = \omega(\pi(\sigma^{-1}))$ belongs to $K[G]$ and, for $\tau \in H$, $p(\sigma\tau) = p(\sigma)$ so that $p \in K[G]^H$. Hence $p = \theta \circ \pi$ for some $\theta \in K[S]$. However $\theta : S \to K$ is the inversion mapping $S \to S$ followed by $\omega : S \to K$. Consequently the inversion mapping on S is a K-morphism and now the proof is complete.

We have now progressed a considerable way towards a satisfactory definition of a quotient group of an affine group. The next theorem provides the main ingredient that is missing.

Theorem 47. *Let G and Γ be connected affine groups defined ov[er] an algebraically closed field K and let $\pi : G \to \Gamma$ be a surjective K-homomorphism. Then π is an open mapping.*

Proof. This will be divided into two stages. Put $H = \operatorname{Ker} \pi$ so tha[t] H is a closed, normal subgroup of G.

First stage. Here it will be assumed that H *itself is connected.* We set $R = K[G]$ and $I = K[\Gamma]$. These are integral domains, I is a K-subalgebra of R, and there exist elements $u_1, u_2, \ldots, u_m$ in R such that $R = I[u_1, u_2, \ldots, u_m]$, that is to say R is a finitely generated

I-algebra.

By Theorem 13, all the points of G are simple and therefore, as we remarked earlier, the local rings of the points of G are unique factorization domains. Moreover the intersection of these unique factorization domains is $K[G] = R$. Consequently R is integrally closed and, for similar reasons, I is integrally closed as well.

There exist† $a \in I$, $a \neq 0$ and $z_1, z_2, \ldots, z_q$ in R so that $z_1, z_2, \ldots, z_q$ are algebraically independent over I and $R[a^{-1}]$ is an integral extension of $I[a^{-1}, z_1, \ldots, z_q]$. We note that $R[a^{-1}]$, $I[z_1, z_2, \ldots, z_q]$ and $I[a^{-1}, z_1, \ldots, z_q]$ are all of them integrally closed.‡

Choose a maximal ideal M, of R, so that $a \notin M$ and let x be the corresponding point of G. Then $y = \pi(x)$ corresponds to the maximal ideal $N = M \cap I$ of $I = K[\Gamma]$.

Let U be an open neighbourhood of x in G. We shall prove that $\pi(U)$ is a neighbourhood of y. Since π is a surjective homomorphism, the fact that π is open will then follow by homogeneity. Before proceeding note there exists $\omega \in R$ such that $\omega(x) \neq 0$ and

$$\{\xi \mid \xi \in G \text{ and } \omega(\xi) \neq 0\} \subseteq U.$$

It follows that we may assume that U itself has the form $\{\xi \mid \xi \in G \text{ and } \omega(\xi) \neq 0\}$ for a suitably chosen ω.

We have $\pi^{-1}(\{y\}) = Hx$ and this is irreducible because we are assuming that H is connected. It follows that NR has exactly one minimal prime ideal and, because $\omega(x) \neq 0$, ω does not belong to this minimal prime ideal. But $a \notin M$ and therefore $a \notin N$. We now see that $NR[a^{-1}]$ has precisely one minimal prime ideal and ω does not belong to it.

Since $z_1, z_2, \ldots, z_q$ are algebraically independent over I, $NI[z_1, z_2, \ldots, z_q]$ is a prime ideal of $I[z_1, z_2, \ldots, z_q]$ which does not contain a. Consequently $NI[a^{-1}, z_1, \ldots, z_q]$ is a prime ideal of $I[a^{-1}, z_1, \ldots, z_q]$. Furthermore, because $R[a^{-1}]$ is an integral ex-

---

† See [(8) Theorem 14.4, p. 45].

‡ Note that if R is an integrally closed integral domain, then so is the polynomial ring $R[X]$. See [(13) Vol. 2, Theorem 29, p. 85].

tension of $I[a^{-1}, z_1, \ldots, z_q]$, every prime ideal of $R[a^{-1}]$ that contracts to $NI[a^{-1}, z_1, \ldots, z_q]$ in $I[a^{-1}, z_1, \ldots, z_q]$ has to be a minimal prime ideal of $NR[a^{-1}]$. Thus we reach the following conclusion: <u>there is only one prime ideal of</u> $R[a^{-1}]$ <u>that contracts to</u> $NI[a^{-1}, z_1, \ldots, z_q]$ <u>and it does not contain</u> ω.

Since $\omega \in R$ it is integral over the integrally closed ring $I[a^{-1}, z_1, \ldots, z_q]$, and therefore the irreducible polynomial for ω over the quotient field of $I[a^{-1}, z_1, \ldots, z_q]$ has the form

$$X^h + \frac{\phi_1(z_1, \ldots, z_q)}{a^t} X^{h-1} + \ldots + \frac{\phi_h(z_1, \ldots, z_q)}{a^t},$$

where $t \geq 0$ and $\phi_i(z_1, z_2, \ldots, z_q) \in I[z_1, z_2, \ldots, z_q]$. Also $\phi_h(z_1, z_2, \ldots, z_q) \notin NI[z_1, z_2, \ldots, z_q]$ because ω is not in any prime ideal of $R[a^{-1}]$ that contracts to $NI[a^{-1}, z_1, \ldots, z_q]$. Let v be a coefficient of $\phi_h(z_1, z_2, \ldots, z_q)$ that is not in N. Then $av \in K[\Gamma]$ and $av \notin N$. Accordingly

$$T = \{\eta \mid \eta \in \Gamma \text{ and } a(\eta)v(\eta) \neq 0\}$$

is open in Γ and it contains y.

<u>We assert that</u> $T \subseteq \pi(U)$. For let N' be any maximal ideal of I that corresponds to a point of T. Then $av \notin N'$. Choose $\alpha_1, \alpha_2, \ldots, \alpha_q$ in K so that $\phi_h(\alpha_1, \alpha_2, \ldots, \alpha_q) \notin N'$. (This is possible because $v \notin N'$.) The existence of the α_i shows that there is a maximal ideal of $I[z_1, z_2, \ldots, z_q]$ that contracts to N' in I, but contains neither a nor $\phi_h(z_1, z_2, \ldots, z_q)$. Consequently there is a maximal ideal of $I[a^{-1}, z_1, \ldots, z_q]$ that contracts to N' but does not contain $\phi_h(z_1, z_2, \ldots, z_q)/a^t$. Now

$$\omega^h + \frac{\phi_1(z_1, \ldots, z_q)}{a^t} \omega^{h-1} + \ldots + \frac{\phi_h(z_1, \ldots, z_q)}{a^t} = 0$$

and $R[a^{-1}]$ is an integral extension of $I[a^{-1}, z_1, \ldots, z_q]$. Accordingly there is a maximal ideal of $R[a^{-1}]$ that contracts to N' but does not contain ω. Finally we see there is a maximal ideal M', of R, such that $M' \cap I = N'$ and $\omega \notin M'$. Thus if N' corresponds to the point

$y' \in T$ and M' to $x' \in G$, then $x' \in U$ and $y' = \pi(x')$. This proves that $T \subseteq \pi(U)$ and establishes that π is open for the special case where H is connected.

Second stage. Here we remove our assumption concerning H. To this end let H_0 be the connected component of the identity of H. If $x \in G$, then xH_0x^{-1} is a closed connected subgroup of $xHx^{-1} = H$. Consequently $xH_0x^{-1} \subseteq H_0$ and therefore H_0 is a closed, connected, normal subgroup of G. By Lemmas 24 and 25, the group G/H_0 can be given an affine structure in such a way that (i) the natural mapping $\theta : G \to G/H_0$ is a K-homomorphism, and (ii) with the obvious identification

$$K[G/H_0] = K[G]^{H_0}.$$

Also, in view of what was established in the first stage of the present proof, θ is an open mapping.

Consider the finite group H/H_0. This is a normal subgroup of G/H_0 and its finiteness ensures that it is a closed subgroup. Let us regard H/H_0 as acting morphically on the right of G/H_0. By Theorem 45, we can find a strict quotient, $\chi : G/H_0 \to S$ say, for this action and then we have

$$K[S] = K[G/H_0]^{(H/H_0)} = (K[G]^{H_0})^{(H/H_0)} = K[G]^H.$$

Also, by Lemma 25, we can impose a group structure on S so as to turn it into an affine group and χ into a K-homomorphism. Put $\rho = \chi \circ \theta$. Then $\rho : G \to S$ is an open, surjective K-homomorphism with $H = \operatorname{Ker} \rho$ and $K[S] = K[G]^H$.

Since $K[\Gamma] \subseteq K[S]$ (see (5.13.1)), there is a K-morphism $\lambda : S \to \Gamma$, of affine sets, such that $\lambda \circ \rho = \pi$. To complete the proof we have only to show that λ is open.

If the characteristic of K is zero, then (as we saw in the proof of Lemma 23) $K[\Gamma] = K[G]^H = K[S]$ so λ is an isomorphism and there is no problem. Suppose therefore that the characteristic of K is the prime p and let $f \in K[G]^H$. Then (again by the proof of Lemma 23) $f^{p^n} \in K[\Gamma]$ for a suitable positive integer n. In fact $K[S]$ is an integral extension of $K[\Gamma]$ and the maximal ideals of $K[S]$ that do not contain f have as their contractions precisely the maximal ideals of $K[\Gamma]$ that do not contain f^{p^n}. This ensures that $\lambda : S \to \Gamma$ is an open mapping.

We come now to the main result of this section.

Theorem 48. _Let_ G _be an affine group defined over an algebraically closed field_ K _and let_ H _be a closed, normal subgroup of_ G. _Then there exists an affine group_ S _and an open, surjective_ K-_homomorphism_ $\pi : G \to S$ _such that_ $\operatorname{Ker} \pi = H$ _and, as a subalgebra of_ $K[G]$, $K[S] = K[G]^H$

Proof. Let G_0 be the connected component of the identity of G and let $G_0, G_1, \ldots, G_r$ be the different cosets of G_0 in G. These are disjoint, closed, irreducible subsets of G.

Assume, for the moment, that $H \subseteq G_0$. By combining Lemmas 24 and 25 with Theorem 47, we see that there exists a strict quotient $\pi_0 : G_0 \to S_0$ for the right action of H on G_0.

Choose $\sigma_i \in G$ so that $G_i = \sigma_i G_0$. Then left multiplication by σ_i induces a K-isomorphism $\phi_i : G_0 \xrightarrow{\sim} G_i$. Now H acts morphically on the right of both G_0 and G_i, and ϕ_i commutes with the action of the elements of H. It follows that there exists a strict quotient $\pi_i : G_i \to S_i$ for the action of H on G_i.

Let S be the disjoint union of $S_0, S_1, \ldots, S_r$. We know that S can be regarded as an affine set with

$$K[S] = K[G_0]^H \times K[G_1]^H \times \ldots \times K[G_r]^H.$$

Also each $\tau \in H$ induces a K-algebra automorphism of $K[G_i]$ $(i = 0, 1, \ldots, r)$ and thereby a K-algebra automorphism of their direct product. On this understanding

$$K[S] = (K[G_0] \times K[G_1] \times \ldots \times K[G_r])^H.$$

By Lemma 19,

$$K[G] = K[G_0] \times K[G_1] \times \ldots \times K[G_r]$$

whence $K[S] = K[G]^H$. Moreover the inclusion mapping of $K[G]^H$ in $K[G]$ gives rise to a K-morphism $\pi : G \to S$ which extends the various morphisms $\pi_i : G_i \to S_i$ (see the discussion following the proof of Lemma 19). It is now clear that $\pi : G \to S$ is a strict quotient for the action of H on G, and that if we use Lemma 25 to turn S into an affine group, then $\pi : G \to S$ meets the requirements of the theorem.

It remains for us to remove the assumption that H is contained in G_0. To this end let H_0 denote the connected component of the identity of H. Then H_0 is not only a closed, normal subgroup of H, but also† a closed normal subgroup of G. Because $H_0 \subseteq G_0$, the first part of this proof now shows that we can regard the abstract group G/H_0 as an affine group; moreover this can be done in such a way that the natural mapping $\theta : G \to G/H_0$ is an open K-homomorphism allowing us to identify $K[G/H_0]$ with $K[G]^{H_0}$. Next the finite group H/H_0 is a closed, normal subgroup of G/H_0. As in the proof of Theorem 47, we can now find an affine group S and an open surjective K-homomorphism $\chi : G/H_0 \to S$ such that (i) Ker $\chi = H/H_0$ and (ii) K[S], as a subalgebra of $K[G/H_0]$, is

$$K[G/H_0]^{H/H_0} = (K[G]^{H_0})^{H/H_0} = (K[G])^H.$$

If therefore we put $\pi = \chi \circ \theta$, then $\pi : G \to S$ is a K-homomorphism with all the required properties.

† See the second part of the proof of Theorem 47.

6 · The associated Lie algebra

General remarks

As usual K denotes an arbitrary field. In this chapter we continue with the study of affine groups defined over K. If G is such a group then e (or e_G) is used to denote the identity element of G, and G_0 the connected component containing e. For $f \in K[G]$ and $\sigma \in G$, the notation f^σ is always employed to describe that member of K[G] which satisfies $f^\sigma(\tau) = f(\tau\sigma)$ for all $\tau \in G$. Consequently the formulae (5.5.5) (and not (5.5.4)) will be applicable.

The associated Lie algebra (see section (6.3)) of the affine group G is denoted by $\underset{\sim}{g}$ and this notation is adapted in a natural way to deal with special situations. For example, the Lie algebra of GL(V) is written as $\underset{\sim\sim}{gl}(V)$.

6.1 General K-algebras†

So far the K-algebras that have concerned us have all been associative. It is now necessary to consider a wider class of algebras.

A general K-algebra is a pair (M, μ), where M is a vector space over K and

$$\mu : M \times M \to M \tag{6.1.1}$$

is a bilinear mapping of $M \times M$ into M. We shall call $\mu(x, y)$, where $x, y \in M$, the product of x and y. However multiplication is no longer assumed to be associative.

Let (M, μ) be a general K-algebra and suppose that N is a subspace of M with the property that $\mu(x, y) \in N$ whenever $x, y \in N$. In

† General algebras are more commonly called non-associative algebras.

these circumstances μ induces a bilinear mapping

$$\bar{\mu} : N \times N \to N$$

thereby making $(N, \bar{\mu})$ a general K-algebra. We call $(N, \bar{\mu})$ a <u>sub-algebra</u> of (M, μ).

To see how general K-algebras can arise, let M be any vector space over K and $\{m_i\}_{i \in I}$ a base for M over K. Further let $\{\xi_{ij}\}$, where (i, j) varies freely in $I \times I$, be an arbitrary family of elements of M. Then there is precisely one way to turn M into a general K-algebra so that the product of m_i and m_j (in that order) is ξ_{ij}.

As an application of this observation assume that (M, μ) is a general K-algebra, let L be an extension field of K, and define the L-space M^L as in section (1.6). Then M is contained in M^L and there is a unique way to turn M^L into a general L-algebra so that multiplication on M^L extends multiplication on M.

Suppose next that (M, μ) and (M', μ') are general K-algebras. A mapping $f : M \to M'$ is called a <u>homomorphism</u> of general K-algebras if it is K-linear and

$$f(\mu(x, y)) = \mu'(f(x), f(y))$$

for all $x, y \in M$. In appropriate circumstances two such homomorphisms can be combined to give a new homomorphism.

Let $f : M \to M'$ be a homomorphism of general K-algebras. If f is a bijection, then it is said to be an <u>isomorphism.</u> In such a case $f^{-1} : M' \to M$ is also an isomorphism and M and M' are said to be <u>isomorphic.</u> The identity mapping of M is, of course, an isomorphism of general K-algebras.

Finally let (M_1, μ_1) and (M_2, μ_2) be general K-algebras and denote by M the direct sum of the K-spaces M_1 and M_2. Then the mapping

$$\mu : M \times M \to M$$

given by

$$\mu((x, y), (x', y')) = (\mu_1(x, x'), \mu_2(y, y'))$$

is bilinear and therefore (M, μ) is a general K-algebra. We shall call it the direct product of M_1 and M_2 and denote it by $M_1 \times M_2$.

Before we leave this section it should be noted that, with two exceptions, the terminology used above is compatible with that employed in Chapter 1 in connection with associative algebras. The exceptions occur in connection with unitary K-algebras. Here subalgebras are always required to share the identity element of the ambient algebra and homomorphisms have to preserve identity elements.

6.2 Lie algebras

Let (M, μ) be a general K-algebra in the sense of section (6.1).

Definition. The K-algebra (M, μ) is said to be a 'Lie algebra' provided that

(i) $\mu(x, x) = 0$ for all $x \in M$;

(ii) $\mu(\mu(x, y), z) + \mu(\mu(y, z), x) + \mu(\mu(z, x), y) = 0$ whenever x, y, z are in M.

Condition (ii) is known as Jacobi's identity.

Obviously if two general K-algebras are isomorphic and one is a Lie algebra, then so is the other. Every subalgebra of a Lie algebra is a Lie algebra. Any vector space can be turned into a Lie algebra by defining the product $\mu(x, y)$ to be zero[†] for all choices of x and y.

Theorem 1. *Let* (M, μ) *be a Lie algebra. Then*

$$\mu(x, y) + \mu(y, x) = 0$$

for all $x, y \in M$.

Proof. We have $\mu(x+y, x+y) = 0$. The desired result follows by expanding the left hand side and using the fact that $\mu(x, x)$ and $\mu(y, y)$ are both zero.

---

† A Lie algebra in which all products are zero is said to be abelian.

Let (M, μ) be a general K-algebra and $\{e_i\}_{i \in I}$ a base for M over K. Evidently (M, μ) is a Lie algebra if and only if the following three conditions are satisfied:

(a) $\mu(e_i, e_i) = 0$ for all i;

(b) $\mu(e_i, e_j) + \mu(e_j, e_i) = 0$ for all i and j;

(c) $\mu(\mu(e_i, e_j), e_k) + \mu(\mu(e_j, e_k), e_i) + \mu(\mu(e_k, e_i), e_j) = 0$ for all i, j and k.

Theorem 2. Let (M, μ) be a Lie algebra over K and let L be an extension field of K. Then (M^L, μ^L) is a Lie algebra over L.

Remark. M^L was defined in section (1.6). By μ^L is meant the unique extension of μ which turns M^L into an algebra over L.

Proof. We have only to take a base for M over K and use the criterion described above.

Theorem 3. The direct product of two Lie algebras is again a Lie algebra.

The proof involves no more than a trivial verification.

We add a few words about notation and terminology. If the K-space M is provided with a multiplication which makes it a Lie algebra, then the usual notation for the product of x and y is $[x, y]$. We call $[x, y]$ the Lie product of x and y. Hence in addition to being bilinear the Lie product has the following properties:

$$\left.\begin{aligned} &[x, x] = 0 \\ &[x, y] + [y, x] = 0, \\ &[[x, y], z] + [[y, z], x] + [[z, x], y] = 0. \end{aligned}\right\} \qquad (6.2.1)$$

So far as terminology is concerned, it is customary to refer to M itself as a Lie algebra and to suppress any direct reference to the product although the latter is an essential part of the structure.

We shall now illustrate the concept of a Lie algebra by means of a few examples.

Example 1. Let A be an associative K-algebra in which the product of the elements x, y of A is written as xy. Then A becomes a Lie algebra if we put

$$[x, y] = xy - yx. \tag{6.2.2}$$

If in the sequel we have occasion to treat an associative K-algebra as a Lie algebra, it is to be understood that the Lie product is defined as in (6.2.2).

Example 2. Let V be any vector space over K. Then $\mathrm{Hom}_K(V, V) = \mathrm{End}_K(V)$ is an associative K-algebra which becomes a Lie algebra if we put

$$[f, g] = fg - gf.$$

Here fg stands for $f \circ g$.

Example 3. Let n be a positive integer and $M_n(K)$ the associative K-algebra formed by all $n \times n$ matrices with entries in K. This becomes a Lie algebra if we put

$$[A, B] = AB - BA$$

for A, B in $M_n(K)$. Note that the matrices with zero trace form a Lie subalgebra.

Example 4. Let R be a unitary and associative K-algebra. We know that $\mathrm{Der}_K(R)$ is a subspace of $\mathrm{Hom}_K(R, R)$ and Example 2 shows that $\mathrm{Hom}_K(R, R)$ has a natural structure as a Lie algebra. Suppose that D_1, D_2 belong to $\mathrm{Der}_K(R)$ and that $a, b \in R$. Then

$$(D_1D_2)(ab) = (D_1D_2a)b + (D_2a)(D_1b) + (D_1a)(D_2b) + a(D_1D_2b)$$

whence $[D_1, D_2] = D_1D_2 - D_2D_1$ belongs to $\mathrm{Der}_K(R)$. This proves

Theorem 4. _Let R be a unitary and associative K-algebra and let $\mathrm{Hom}_K(R, R)$ be considered as a Lie algebra as in Example 2. Then $\mathrm{Der}_K(R)$ is a Lie subalgebra of $\mathrm{Hom}_K(R, R)$._

We return to the general theory in order to introduce some extra terminology. Let M be a Lie algebra over K. By the dimension of M we mean simply its dimension as a K-space. Next let $x \in M$ and define $\phi_x : M \to M$ by

$$\phi_x(z) = [x, z]. \quad (6.2.3)$$

Then $\phi_x \in \mathrm{End}_K(M)$ and there results a K-linear mapping

$$\mathrm{ad} : M \to \mathrm{End}_K(M) \quad (6.2.4)$$

in which $x \mapsto \phi_x$. An easy verification using Jacobi's identity now shows that

$$\phi_{[x, y]} = \phi_x\phi_y - \phi_y\phi_x.$$

Thus (6.2.4) is a homomorphism of Lie algebras. This homomorphism is called the adjoint representation of M.

6.3 The Lie algebra of an affine group.

Throughout section (6.3) G will denote an affine group defined over K and G_0 the connected component of its identity element e. Let $\sigma \in G$ and $f \in K[G]$. The right translation of G by means of σ induces an automorphism of the K-algebra K[G] and the image of f under the automorphism will be denoted by f^σ. Thus the formulae (5.5.5) are applicable and, in particular,

$$f^\sigma(\tau) = f(\tau\sigma) \quad (6.3.1)$$

for all $\tau \in G$.

Let $D \in \mathrm{Der}_K(K[G])$.

Definition. We say that D is an 'invariant derivation'† if $D(f^\sigma) = (Df)^\sigma$ for all $\sigma \in G$ and $f \in K[G]$.

---

† More precisely D is a right invariant derivation. Left invariant derivations are obtained by making the obvious changes.

Denote by $\mathrm{Der}_K(K[G])^G$ the set of all invariant derivations. By Theorem 4, $\mathrm{Der}_K(K[G])$ is a Lie algebra and now a trivial verification establishes

Theorem 5. *The space* $\mathrm{Der}_K(K[G])^G$ *of invariant derivations is a Lie subalgebra of* $\mathrm{Der}_K(K[G])$.

Definition. The Lie algebra $\mathrm{Der}_K(K[G])^G$ will be called the 'Lie algebra of G'.

It will next be shown that there is an intimate connection between the Lie algebra of G and the tangent space $\mathrm{Der}_K(G, e)$ to G at e.

To this end let $D \in \mathrm{Der}_K(K[G])$, define

$$D_e : K[G] \to K \tag{6.3.2}$$

by

$$D_e f = (Df)(e), \tag{6.3.3}$$

and note that D_e belongs to $\mathrm{Der}_K(K[G], K, \omega_e)$, where $\omega_e : K[G] \to K$ maps f into $f(e)$. Thus $D_e \in \mathrm{Der}_K(G, e)$ and we obtain a K-linear mapping

$$\mathrm{Der}_K(K[G]) \to \mathrm{Der}_K(G, e)$$

in which $D \mapsto D_e$. This in turn induces a K-linear mapping

$$\mathrm{Der}_K(K[G])^G \to \mathrm{Der}_K(G, e). \tag{6.3.4}$$

Theorem 6. *The mapping*

$$\mathrm{Der}_K(K[G])^G \to \mathrm{Der}_K(G, e)$$

of (6.3.4) is an isomorphism of K-spaces.

Proof. First suppose that $D \in \mathrm{Der}_K(K[G])^G$ and $D_e = 0$. If $\sigma \in G$ and $f \in K[G]$, then

$$(Df)(\sigma) = (Df)^{\sigma}(e) = (Df^{\sigma})(e) = D_e f^{\sigma} = 0.$$

It follows that $Df = 0$ for all f in $K[G]$ and hence that $D = 0$. Thus (6.3.4) is an injection.

Now assume that $\Delta \in \mathrm{Der}_K(G, e)$. For each $f \in K[G]$ we define a mapping

$$Df : G \to K$$

by $(Df)(\sigma) = \Delta(f^\sigma)$. Now the mapping $G \times G \to K$ in which $(\sigma, \tau) \mapsto f(\sigma\tau)$ belongs to $K[G \times G]$. Consequently there exist $f_1, f_2, \ldots, f_m$ and $g_1, g_2, \ldots, g_m$ in $K[G]$ such that

$$f(\sigma\tau) = \sum_{i=1}^{m} f_i(\sigma)g_i(\tau)$$

for all σ, τ in G. Since $f^\tau(\sigma) = f(\sigma\tau)$ it follows that

$$f^\tau = \sum_{i=1}^{m} g_i(\tau)f_i$$

and hence that

$$(Df)(\tau) = \sum_{i=1}^{m} g_i(\tau)\Delta f_i .$$

Accordingly

$$Df = \sum_{i=1}^{m} (\Delta f_i)g_i$$

and thus we see that $Df \in K[G]$. Consequently $f \mapsto Df$ provides a K-linear mapping

$$D : K[G] \to K[G].$$

Suppose next that $f, g \in K[G]$. Then

$$\begin{aligned}(D(fg))(\sigma) &= \Delta(f^\sigma g^\sigma)\\ &= g^\sigma(e)((Df)(\sigma)) + f^\sigma(e)((Dg)(\sigma))\\ &= g(\sigma)((Df)(\sigma)) + f(\sigma)((Dg)(\sigma))\\ &= (g(Df) + f(Dg))(\sigma)\end{aligned}$$

and this shows that $D \in \mathrm{Der}_K(K[G])$. Obviously $D_e = \Delta$. Finally, for $\sigma, \tau \in G$ and $f \in K[G]$, we have

$$
\begin{aligned}
(Df^{\tau})(\sigma) &= \Delta((f^{\tau})^{\sigma}) \\
&= \Delta(f^{\sigma\tau}) \\
&= (Df)(\sigma\tau) \\
&= (Df)^{\tau}(\sigma)
\end{aligned}
$$

which shows that $D \in \mathrm{Der}_K(K[G])^G$. This establishes that (6.3.4) is a surjection and completes the proof.

Corollary. _Let_ $D, D' \in \mathrm{Der}_K(K[G])^G$. _Then_ $D = D'$ _if and only if_ $D_e = D'_e$.

It will now be shown how a homomorphism of affine groups leads to a homomorphism of their Lie algebras. To this end let $\phi : G \to G'$ be a K-homomorphism of affine groups. By Theorem 6, we have isomorphisms

$$\mathrm{Der}_K(K[G])^G \xrightarrow{\sim} \mathrm{Der}_K(G, e)$$

and

$$\mathrm{Der}_K(K[G'])^{G'} \xrightarrow{\sim} \mathrm{Der}_K(G', e')$$

of K-spaces, and, by Chapter 4 Theorem 16, we have a K-linear mapping

$$d(\phi, e) : \mathrm{Der}_K(G, e) \to \mathrm{Der}_K(G', e').$$

(Here $e' = \phi(e)$ is the identity element of G'.) We can therefore define, in a unique manner, a K-linear mapping

$$d\phi : \mathrm{Der}_K(K[G])^G \to \mathrm{Der}_K(K[G'])^{G'} \tag{6.3.5}$$

so as to make

$$
\begin{array}{ccc}
\mathrm{Der}_K(K[G])^G & \xrightarrow{\sim} & \mathrm{Der}_K(G, e) \\
{\scriptstyle d\phi}\downarrow & & \downarrow{\scriptstyle d(\phi, e)} \\
\mathrm{Der}_K(K[G'])^{G'} & \xrightarrow{\sim} & \mathrm{Der}_K(G', e')
\end{array}
\tag{6.3.6}
$$

a commutative diagram. Thus if $D \in \mathrm{Der}_K(K[G])^G$, then

$$(d\phi D)_{e'} = D_e \circ \phi^*,$$

where $\phi^* : K[G'] \to K[G]$ is the homomorphism of K-algebras induced by ϕ, and therefore

$$((d\phi D)f')(e') = (D(f' \circ \phi))(e) \tag{6.3.7}$$

for all $f' \in K[G']$.

Lemma 1. _Suppose that_ $\sigma \in G$, $f' \in K[G']$ _and_ $D \in \mathrm{Der}_K(K[G])^G$. _Then_

(a) $(f' \circ \phi)^\sigma = f'^{\phi(\sigma)} \circ \phi$;

(b) $(d\phi D)f' \circ \phi = D(f' \circ \phi)$.

Proof. Let $\tau \in G$. Then

$$\begin{aligned}(f' \circ \phi)^\sigma(\tau) &= f'(\phi(\tau\sigma)) \\ &= f'(\phi(\tau)\phi(\sigma)) \\ &= (f'^{\phi(\sigma)} \circ \phi)(\tau).\end{aligned}$$

This proves (a). Next

$$\begin{aligned}((d\phi D)f' \circ \phi)(\sigma) &= ((d\phi D)f')^{\phi(\sigma)}(e') \\ &= ((d\phi D)f'^{\phi(\sigma)})(e') \\ &= (d\phi D)_{e'}(f'^{\phi(\sigma)}) \\ &= D_e(f'^{\phi(\sigma)} \circ \phi)\end{aligned}$$

by (6.3.7). Hence, using (a),

$$\begin{aligned}((d\phi D)f' \circ \phi)(\sigma) &= (D(f' \circ \phi)^\sigma)(e) \\ &= (D(f' \circ \phi))(\sigma)\end{aligned}$$

and (b) is established as well.

We are now ready to prove the important

Theorem 7. _Let_ $\phi : G \to G'$ _be a_ K-_homomorphism of affine groups. Then_

$$d\phi : \mathrm{Der}_K(K[G])^G \to \mathrm{Der}_K(K[G'])^{G'}$$

is a homomorphism of Lie algebras.

Proof. We use the following notation. When $D \in \mathrm{Der}_K(K[G])^G$ we put $\overline{D} = d\phi D$ and $\Delta = D_e$. Thus when $f' \in K[G']$ we have, from (6. 3. 7),

$$\overline{D}_{e'} f' = \Delta(f' \circ \phi). \tag{6. 3. 8}$$

Now let $D^{(1)}, D^{(2)}$ belong to $\mathrm{Der}_K(K[G])^G$. Then

$$\begin{aligned}([\overline{D}^{(1)}, \overline{D}^{(2)}]f')(e') &= (\overline{D}^{(1)}\overline{D}^{(2)}f' - \overline{D}^{(2)}\overline{D}^{(1)}f')(e') \\ &= \overline{D}^{(1)}_{e'}(\overline{D}^{(2)}f') - \overline{D}^{(2)}_{e'}(\overline{D}^{(1)}f')\end{aligned}$$

and hence

$$[\overline{D}^{(1)}, \overline{D}^{(2)}]_{e'} f' = \Delta^{(1)}(\overline{D}^{(2)}f' \circ \phi) - \Delta^{(2)}(\overline{D}^{(1)}f' \circ \phi)$$

by (6. 3. 8). On the other hand (6. 3. 8) also shows that

$$\begin{aligned}\left(\overline{[D^{(1)}, D^{(2)}]}\right)_{e'} f' &= [D^{(1)}, D^{(2)}]_e (f' \circ \phi) \\ &= (D^{(1)}D^{(2)}(f' \circ \phi))(e) - (D^{(2)}D^{(1)}(f' \circ \phi))(e) \\ &= \Delta^{(1)}(D^{(2)}(f' \circ \phi)) - \Delta^{(2)}(D^{(1)}(f' \circ \phi)) \\ &= \Delta^{(1)}(\overline{D}^{(2)}f' \circ \phi) - \Delta^{(2)}(\overline{D}^{(1)}f' \circ \phi)\end{aligned}$$

by (b) of Lemma 1. It therefore follows that

$$\left(\overline{[D^{(1)}, D^{(2)}]}\right)_{e'} f' = [\overline{D}^{(1)}, \overline{D}^{(2)}]_{e'} f'.$$

Consequently $\overline{([D^{(1)}, D^{(2)}])}_{e'} = [\overline{D}^{(1)}, \overline{D}^{(2)}]_{e'}$ whence

$$\overline{[D^{(1)}, D^{(2)}]} = [\overline{D}^{(1)}, \overline{D}^{(2)}]$$

by Theorem 6 Cor. and this completes the proof.

Theorem 8. Let $\phi : G \to G'$ and $\psi : G' \to G''$ be K-homomorphisms of affine groups. Then the following hold:

(a) if ϕ is an identity mapping, then so is $d\phi$;

(b) $d(\psi \circ \phi) = d\psi \circ d\phi$;

(c) *if* ϕ *is a* K-*isomorphism of affine groups, then* $d\phi$ *is an isomorphism of Lie groups and* $d(\phi^{-1}) = (d\phi)^{-1}$.

Proof. The first two assertions follow from the properties of the diagram (6.3.6) when taken in conjunction with the remarks that immediately follow the statement of Theorem 16 of Chapter 4. The final assertion is a consequence of the first two.

We next investigate the connection between the Lie algebra of G and that of a closed subgroup.

Theorem 9. *Let* H *be a closed subgroup of the affine group* G *and let* $\phi : H \to G$ *be the inclusion homomorphism. Then*

$$d\phi : \mathrm{Der}_K(K[H])^H \to \mathrm{Der}_K(K[G])^G$$

is an injection and therefore the Lie algebra of H *can be regarded as a subalgebra of the Lie algebra of* G.

Proof. Let $D \in \mathrm{Der}_K(K[H])^H$ and be such that $d\phi D = 0$, and let $f \in K[G]$. By (6.3.7)

$$D_e(f \circ \phi) = ((d\phi D)f)(e) = 0.$$

But $f \circ \phi$ is a typical member of $K[H]$. Consequently $D_e = 0$ and therefore $D = 0$. The theorem follows.

Theorem 10. *Let* H *be a closed subgroup of the affine group* G *and let the Lie algebra of* H *be regarded as a subalgebra of the Lie algebra of* G. *Further let* D *belong to the Lie algebra of* G. *Then* D *belongs to the Lie algebra of* H *if and only if* $D_e(I_G(H)) = (0)$.

Proof. Let $\phi : H \to G$ be the inclusion homomorphism and let $D' \in \mathrm{Der}_K(K[H])^H$. If now $f \in I_G(H)$, then

$$(d\phi D')_e(f) = D'_e(f \circ \phi) = D'_e 0 = 0.$$

This shows that if D belongs to the Lie algebra of H, then $D_e(I_G(H))$ contains only zero.

Now suppose that $D_e(I_G(H)) = (0)$. The homomorphism $\phi^* : K[G] \to K[H]$ of K-algebras that is induced by ϕ has kernel $I_G(H)$ and therefore there exists $\Delta \in \mathrm{Der}_K(H, e)$ such that

$$D_e = \Delta \circ \phi^*.$$

Next, by Theorem 6, there exists $D' \in \mathrm{Der}_K(K[H])^H$ such that $D'_e = \Delta$. Accordingly

$$(d\phi D')_e = D'_e \circ \phi^* = D_e$$

and now it follows that $d\phi D' = D$. In other terms D belongs to the Lie algebra of H.

This theorem will now be recast in a different form.

Theorem 11. _Let H be a closed subgroup of G and let D belong to the Lie algebra of G. Then D belongs to the Lie algebra of H if and only if $D(I_G(H)) \subseteq I_G(H)$._

Proof. Suppose first of all that $D(I_G(H)) \subseteq I_G(H)$ and let $f \in I_G(H)$. Since $e \in H$ and $Df \in I_G(H)$, we have $(Df)(e) = 0$ and therefore $D_e f = 0$. Thus $D_e(I_G(H)) = 0$ and hence D belongs to the Lie algebra of H by Theorem 10.

Next assume that D is in the Lie algebra of H. Let $f \in I_G(H)$ and let $\tau \in H$. Then $f(H) = \{0\}$ and therefore $f^\tau(H\tau^{-1}) = \{0\}$. But $H\tau^{-1} = H$ and so it follows that $f^\tau \in I_G(H)$. By Theorem 10, $D_e f^\tau = 0$ that is to say $(Df^\tau)(e) = 0$. We now have

$$(Df)(\tau) = (Df)^\tau(e) = (Df^\tau)(e) = 0.$$

This shows that $Df \in I_G(H)$ and hence that $D(I_G(H)) \subseteq I_G(H)$ as required.

Let G_0 be the connected component of the identity of G. Then, by Theorem 9, the Lie algebra of G_0 is contained in the Lie algebra of G. In fact we have

Theorem 12. _The Lie algebra of G_0 coincides with the Lie algebra algebra of G._

Proof. Suppose that D belongs to the Lie algebra of G and let $f \in I_G(G_0)$. By Theorem 10, the desired result will follow if we show that $D_e f$ is zero.

We can choose $h \in K[G]$ so that (i) h vanishes on all the cosets of G_0 in G other than G_0 itself, and (ii) $h(e) \neq 0$. This secures that $fh = 0$ and we also know that $f(e) = 0$. Accordingly

$$0 = D_e(fh) = (D_e f)h(e)$$

and therefore $D_e f = 0$ as required.

We next determine the dimension of the Lie algebra of G.

Theorem 13. The dimension of the Lie algebra of the affine group G is equal to Dim G.

Proof. By Theorem 12 we may suppose that G is connected in which case Chapter 5 Theorem 13 shows that e is a simple point of G. Consequently $\mathrm{Der}_K(G, e)$ is a vector space whose dimension is Dim G and now the desired result follows from Theorem 6.

At this point we insert a few observations that have to do with the case where G is commutative. In this situation the mapping $j : G \to G$ in which $x \mapsto x^{-1}$ is a K-homomorphism. Our aim will be to identify the associated endomorphism of the Lie algebra. First we prove

Theorem 14. Let† $[G, \mu, j, e]$ be an affine group. Then

$$d(j, e) : \mathrm{Der}_K(G, e) \to \mathrm{Der}_K(G, e)$$

maps each element of $\mathrm{Der}_K(G, e)$ into its negative.

Proof. Let $f \in K[G]$ and define $\omega : G \times G \to K$ by $\omega(\sigma, \tau) = f(\sigma\tau^{-1})$. Then $\omega \in K[G \times G]$. Consequently there exist $f_1, f_2, \ldots, f_m$ and $g_1, g_2, \ldots, g_m$ in K[G] with the property that

$$f(\sigma\tau^{-1}) = \sum_{i=1}^{m} f_i(\sigma)g_i(\tau).$$

---

† The notation is the same as that employed in section (5.1).

Note that $\sum_{i=1}^{m} f_i g_i$ not only belongs to $K[G]$ but is, in fact, a constant.

Let $D \in \mathrm{Der}_K(G, e)$. Then

$$\sum_{i=1}^{m} (Df_i)g_i(e) + \sum_{i=1}^{m} f_i(e)(Dg_i) = 0.$$

Now

$$\sum_{i=1}^{m} g_i(e)f_i = f$$

and therefore

$$Df = \sum_{i=1}^{m} (Df_i)g_i(e).$$

On the other hand

$$f \circ j = \sum_{i=1}^{m} f_i(e)g_i$$

and so we obtain

$$(d(j, e)D)f = D(f \circ j) = \sum_{i=1}^{m} f_i(e)(Dg_i).$$

Thus $(d(j, e)D)f = (-D)f$ and the theorem follows.

If G is <u>commutative,</u> then the inversion mapping $j : G \to G$ is a K-automorphism of G and therefore dj is an automorphism of the Lie algebra of G.

Theorem 15. *Let* $[G, \mu, j, e]$ *be an affine group and suppose that* G *is commutative. Then* dj *maps each element of the Lie algebra of* G *into its negative.*

This follows from Theorem 14 and the properties of the diagram (6.3.6).

Still assuming that G is commutative, let D, D' belong to the Lie algebra of G. Since

$$dj[D, D'] = [djD, djD'].$$

We conclude that $-[D, D'] = [D, D']$. Hence if the characteristic of K

is not 2, then $[D, D'] = 0$. However this restriction on the characteristic is not necessary.[†]

We postpone the consideration of examples and continue with the general theory. In what follows G and H denote affine groups defined over K and we put

$$\underset{\approx}{g} = \mathrm{Der}_K(K[G])^G \tag{6.3.9}$$

and

$$\underset{\sim}{h} = \mathrm{Der}_K(K[H])^H. \tag{6.3.10}$$

Thus $\underset{\approx}{g}$ respectively $\underset{\sim}{h}$ is the Lie algebra of G respectively H. This will be standard notation from here on.

A K-homomorphism $\phi : G \to H$ induces a Lie homomorphism $d\phi : \underset{\approx}{g} \to \underset{\sim}{h}$. So far we have no general result that will tell us when $d\phi$ is surjective. It is to this question that we now turn our attention.

Suppose that the K-homomorphism $\phi : G \to H$ is almost surjective and let G_0 respectively H_0 be the connected component of the identity of G respectively H. By Chapter 5 Theorem 23, ϕ induces an almost surjective K-homomorphism $\phi_0 : G_0 \to H_0$. Now $K[G_0]$ is an integral domain and it contains $K[H_0]$ as a subalgebra. Consequently $K(H_0)$ is a subfield of $K(G_0)$. Also G and G_0 have the same Lie algebra $\underset{\approx}{g}$, and the Lie algebra $\underset{\sim}{h}$ of H is also that of H_0. Moreover $d\phi : \underset{\approx}{g} \to \underset{\sim}{h}$ coincides with $d\phi_0 : \underset{\approx}{g} \to \underset{\sim}{h}$.

Definition. The almost surjective K-homomorphism $\phi : G \to H$ is said to be 'separable' if $K(G_0)$ is a separable extension of $K(H_0)$.

Lemma 2. If the almost surjective K-homomorphism $\phi : G \to H$ is separable, then $d\phi : \underset{\approx}{g} \to \underset{\sim}{h}$ is a surjection.

Proof. The preliminary discussion shows that we may assume that G and H are connected. Let $D \in \underset{\sim}{h}$. Then $D \in \mathrm{Der}_K(K[H])^H$ and it extends to a derivation of K(H) over K. This extension will also be denoted by D.

---

† See Theorem 30.

By hypothesis, there exists a field Ω, between $K(H)$ and $K(G)$, such that (i) Ω is a pure transcendental extension of $K(H)$, (ii) $K(G)$ is algebraic and separable over Ω, and (iii) the degree of $K(G)$ over Ω is finite. It is clear that D can be extended to a derivation of Ω over K. By Chapter 4 Theorem 7 it can be extended further to provide a derivation, D' say, of K(G) over K. Thus $D' \in \mathrm{Der}_K(K(G))$ and $D'p = Dp$ for all $p \in K[H]$.

Let $\sigma \in G$. There is an automorphism of the K-algebra $K[G]$ in which $f \mapsto f^\sigma$, and this extends to an automorphism of $K(G)$ over K. We use ξ^σ to denote the image of ξ, where $\xi \in K(G)$, under the extended automorphism.

For the moment we keep σ fixed and define $D'' \in \mathrm{Der}_K(K(G))$ by

$$D''\xi = (D'\xi^{\sigma^{-1}})^\sigma.$$

Then, for $p \in K[H]$ and $\tau \in H$, we have

$$\begin{aligned}(p^{\phi(\sigma)} \circ \phi)(\tau) &= p(\phi(\tau)\phi(\sigma)) \\ &= p(\phi(\tau\sigma)) \\ &= (p \circ \phi)(\tau\sigma) \\ &= (p \circ \phi)^\sigma(\tau).\end{aligned}$$

Thus $p^{\phi(\sigma)} \circ \phi = (p \circ \phi)^\sigma$ and we can simplify this to $p^{\phi(\sigma)} = p^\sigma$ by regarding $K[H]$ as being embedded in $K[G]$. Likewise $p^{\phi(\sigma^{-1})} = p^{\sigma^{-1}}$. Accordingly

$$D''p = (D'p^{\phi(\sigma^{-1})})^\sigma.$$

But

$$D'p^{\phi(\sigma^{-1})} = Dp^{\phi(\sigma^{-1})} = (Dp)^{\phi(\sigma^{-1})} = (Dp)^{\sigma^{-1}},$$

because D is an <u>invariant</u> derivation, and therefore $D''p = Dp$. Thus <u>for any choice of</u> $\sigma \in G$, $D'' \in \mathrm{Der}_K(K(G))$ <u>and it extends to</u> D.

We now impose a condition on σ. Let $K[G] = K[u_1, u_2, \ldots, u_n]$. Then $D'u_i = v_i/g$, where $v_1, v_2, \ldots, v_n$, g are in $K[G]$ and $g \neq 0$. Choose σ so that $g(\sigma) \neq 0$. Then $K[G] = K[u_1^\sigma, u_2^\sigma, \ldots, u_n^\sigma]$ and

$$D''u_i^\sigma = (D'u_i)^\sigma = \frac{v_i^\sigma}{g^\sigma} \in Q_{G,e},$$

because $g^\sigma(e) = g(\sigma)$ is not zero. Accordingly $D''(K[G]) \subseteq Q_{G,e}$ and we can define

$$\Delta : K[G] \to K$$

by $\Delta f = (D''f)(e)$. Clearly $\Delta \in \mathrm{Der}_K(G, e)$. Also† for $p \in K[H]$

$$\Delta p = (D''p)(e) = (Dp)(\varepsilon) = D_\varepsilon p.$$

Thus

$$d(\phi, e) : \mathrm{Der}_K(G, e) \to \mathrm{Der}_K(H, \varepsilon)$$

maps Δ into D_ε. It follows that $d(\phi, e)$ is surjective and therefore $d\phi$ is surjective as well.

Theorem 16. Let $\phi : G \to H$ be a surjective K-homomorphism, let $N = \mathrm{Ker}\ \phi$, and denote by $j : N \to G$ the inclusion homomorphism. If now K is algebraically closed and ϕ is separable, then (with a self-explanatory notation) the sequence

$$0 \to \underset{\sim}{n} \xrightarrow{dj} \underset{\sim}{g} \xrightarrow{d\phi} \underset{\sim}{h} \to 0$$

is exact.

Proof. We know that dj is an injection and Lemma 2 shows that $d\phi$ is a surjection. Also, because $\phi \circ j$ can be factored through a trivial group and the Lie algebra of a trivial group is zero-dimensional, $d\phi \circ dj$ is null. Finally Theorem 24 of Chapter 5 shows that $\mathrm{Dim}\ G = \mathrm{Dim}\ N + \mathrm{Dim}\ H$ and hence the sum of the dimensions of $\underset{\sim}{n}$ and $\underset{\sim}{h}$ equals that of $\underset{\sim}{g}$. The theorem follows.

We next examine the Lie algebra of the direct product of two affine groups G and H. To this end suppose that $D \in \underset{\sim}{g}$ and $D' \in \underset{\sim}{h}$. Since $K[G \times H]$ can be identified with $K[G] \otimes_K K[H]$, it follows that there is a K-linear mapping

---

† From here on ε denotes the identity element of H.

$$\Delta_{D,D'} : K[G \times H] \to K[G \times H]$$

in which

$$\Delta_{D,D'}(f \vee g) = Df \vee g + f \vee D'g. \tag{6.3.11}$$

(Here $f \in K[G]$, $g \in K[H]$, and $f \vee g$ is defined as in (2.7.1).) Indeed an easy verification shows that $\Delta_{D,D'}$ belongs to $\mathrm{Der}_K(K[G \times H])$. Note that the mapping

$$\mathrm{Der}_K(K[G])^G \oplus \mathrm{Der}_K(K[H])^H \to \mathrm{Der}_K(K[G \times H]) \tag{6.3.12}$$

given by $(D, D') \mapsto \Delta_{D,D'}$ is not only K-linear but also an injection.

We next observe that if $\sigma \in G$ and $\tau \in H$, then

$$(f \vee g)^{(\sigma, \tau)} = f^\sigma \vee g^\tau$$

from which it follows easily that $\Delta_{D,D'}$ belongs to the Lie algebra of $G \times H$. Consequently (6.3.12) gives rise to a K-linear injection

$$\underset{\sim}{g} \times \underset{\sim}{h} \to \mathrm{Der}_K(K[G \times H])^{G\times H}. \tag{6.3.13}$$

However the two terms in (6.3.13) have the same dimension namely $\mathrm{Dim}\, G + \mathrm{Dim}\, H$. Accordingly (6.3.13) is an isomorphism of K-spaces. Now, by Theorem 3, $\underset{\sim}{g} \times \underset{\sim}{h}$ is a Lie algebra and

$$\Delta_{[(D_1, D_1'), (D_2, D_2')]} = [\Delta_{D_1, D_1'}, \Delta_{D_2, D_2'}]$$

as may be verified without difficulty. Thus (6.3.13) is an isomorphism of Lie algebras. These observations are recorded in

Theorem 17. The Lie algebra of the direct product $G \times H$ is naturally isomorphic to the direct product $\underset{\sim}{g} \times \underset{\sim}{h}$ of the corresponding Lie algebras. Let $D \in \underset{\sim}{g}$ and $D' \in \underset{\sim}{h}$. Then (in this isomorphism) (D, D') is matched with the invariant derivation of $K[G \times H]$ which (for $f \in K[G]$ and $g \in K[H]$) maps $f \vee g$ into $Df \vee g + f \vee D'g$.

Thus, when convenient, we may regard $\underset{\sim}{g} \times \underset{\sim}{h}$ as being the Lie algebra of $G \times H$.

Let G be an affine group defined over K and L an extension field of K. Then G has a Lie algebra $\underset{\sim}{g}$ and $G^{(L)}$ a Lie algebra $\underset{\sim}{g}^*$ say. By Theorem 2, $\underset{\sim}{g}^L$ is a Lie algebra. Our aim is to investigate the connection between $\underset{\sim}{g}^L$ and $\underset{\sim}{g}^*$. First we show that there is a natural embedding of $\underset{\sim}{g}$ in $\underset{\sim}{g}^*$.

Suppose that $D \in \underset{\sim}{g}$. Since D is a K-linear mapping of K[G] into itself it has a unique extension to an L-linear mapping D^L of $K[G]^L = L[G^{(L)}]$ into itself. We wish to show that D^L is in $\underset{\sim}{g}^*$. To this end put $\Delta = D_e$. Then $\Delta \in Der_K(G, e)$ and it has a unique extension Δ^L to an L-linear mapping of $K[G]^L = L[G^{(L)}]$ into $K^L = L$. Further it has already been shown† that $\Delta^L \in Der_L(G^{(L)}, e)$. Next, by Theorem 6, there exists a unique $\overline{D} \in \underset{\sim}{g}^*$ such that $\overline{D}_e = \Delta^L$.

<u>We claim that</u> $\overline{D} = D^L$. To prove this it will suffice to show that they agree on K[G]. Assume therefore that $f \in K[G]$ and let us use $\hat{f}$ to denote that member of $L[G^{(L)}]$ which is its natural prolongation. Then

$$(\overline{D}\hat{f})(e) = \Delta^L\hat{f} = \Delta f = (Df)(e).$$

If now $\sigma \in G$, then $\hat{f}^\sigma$ is the natural prolongation of f^σ, so we may replace f and $\hat{f}$ by f^σ and $\hat{f}^\sigma$ respectively. Since $\overline{D}$ and D are <u>invariant</u> derivations this leads to

$$(\overline{D}\hat{f})(\sigma) = (Df)(\sigma).$$

Accordingly $\overline{D}\hat{f}$ is the natural prolongation of Df and our claim is established. It follows that $D^L \in \underset{\sim}{g}^*$. We therefore have a mapping

$$\underset{\sim}{g} \rightarrow \underset{\sim}{g}^*, \tag{6.4.1}$$

given by $D \mapsto D^L$ which is K-linear and an injection, and which enables us to regard $\underset{\sim}{g}$ as being embedded in $\underset{\sim}{g}^*$. Before proceeding we record some of these observations in

† See section (4.4) and, in particular, the remark just before Theorem 26 in that section.

Theorem 18. Let $\underset{\sim}{g}$ be the Lie algebra of G and let L be an extension field of K. If now $D \in \underset{\sim}{g}$, then D^L (see above) belongs to the Lie algebra of $G^{(L)}$. Furthermore $(D^L)_e = (D_e)^L$.

Now let $D_1, D_2, \ldots, D_p$ be a K-base for $\underset{\sim}{g}$ and put $\Delta_i = (D_i)_e$. By Theorem 6, $\Delta_1, \Delta_2, \ldots, \Delta_p$ is a K-base for $\mathrm{Der}_K(G, e)$ and therefore, by Chapter 4 Theorem 26, $\Delta_1^L, \Delta_2^L, \ldots, \Delta_p^L$ is an L-base for $\mathrm{Der}_L(G^{(L)}, e)$. But we know that $(D_i^L)_e = \Delta_i^L$. It follows, by Theorem 6 that $D_1^L, D_2^L, \ldots, D_p^L$ is an L-base for $\underset{\sim}{g}^*$. Hence when $\underset{\sim}{g}$ is regarded as a K-subspace of $\underset{\sim}{g}^*$ we have $\underset{\sim}{g}^* = \underset{\sim}{g}^L$, where each side is regarded as an L-space. However

$$[D, D']^L = [D^L, D'^L]$$

for all D, D' in $\underset{\sim}{g}$ and thus we see that $\underset{\sim}{g}^*$ and $\underset{\sim}{g}^L$ coincide as Lie algebras. This establishes

Theorem 19. Let $\underset{\sim}{g}$ be the Lie algebra of G and let L be an extension of the ground field K. Then $\underset{\sim}{g}^L$ is the Lie algebra of $G^{(L)}$. (The embedding of $\underset{\sim}{g}$ in the Lie algebra of $G^{(L)}$ is obtained by mapping D, of $\underset{\sim}{g}$, into D^L.)

Let us now examine the effect of enlarging the ground field on the Lie homomorphism $d\phi : \underset{\sim}{h} \to \underset{\sim}{g}$ obtained from a K-homomorphism $\phi : H \to G$. To this end suppose that $D \in \underset{\sim}{h}$, $f \in K[G]$ and, as before, let $\hat{f}$ denote the member of $L[G^{(L)}]$ that prolongs f. From ϕ we obtain an L-homomorphism $\phi^{(L)} : H^{(L)} \to G^{(L)}$ and, from (6.3.7), we have†

$$((d\phi^{(L)} D^L)\hat{f})(e) = (D^L(\hat{f} \circ \phi^{(L)}))(\varepsilon).$$

But $\hat{f} \circ \phi^{(L)}$ is the prolongation of $f \circ \phi$. Consequently

$$((d\phi^{(L)} D^L)\hat{f})(e) = (D(f \circ \phi))(\varepsilon) = ((d\phi D)f)(e)$$

from which we conclude (using the fact that $d\phi D$ and $d\phi^{(L)} D^L$ are invariant derivations) that

---

† We use e respectively ε to denote the identity element of G respectively H.

$$((d\phi^{(L)}D^L)\hat{f})(\sigma) = ((d\phi D)f)(\sigma)$$

for all $\sigma \in G$. Accordingly $(d\phi^{(L)}D^L)\hat{f}$ is the prolongation of $((d\phi)D)f$, that is to say $d\phi^{(L)}D^L$ and $d\phi D$ agree on $K[G]$. Thus $d\phi^{(L)}D^L = (d\phi D)^L$, a result which we now restate as

Theorem 20. <u>Let</u> $\phi : H \to G$ <u>be a K-homomorphism of affine groups and let</u> L <u>be an extension field of</u> K. <u>Then</u>

$$d\phi^{(L)} : \mathfrak{h}^L \to \mathfrak{g}^L$$

<u>extends</u> $d\phi : \mathfrak{h} \to \mathfrak{g}$.

By way of illustration let us consider the case where H is a closed subgroup of G and $\phi : H \to G$ is the inclusion homomorphism. Then $\phi^{(L)} : H^{(L)} \to G^{(L)}$ is also an inclusion homomorphism. As a temporary measure let $\mathfrak{h}^*$ respectively $\mathfrak{g}^*$ denote the Lie algebra of $H^{(L)}$ respectively $G^{(L)}$. We then have injections $\mathfrak{h} \to \mathfrak{g}$, $\mathfrak{h}^* \to \mathfrak{g}^*$, $\mathfrak{h} \to \mathfrak{h}^*$ and $\mathfrak{g} \to \mathfrak{g}^*$, all of which preserve Lie products, and Theorem 20 shows that the diagram

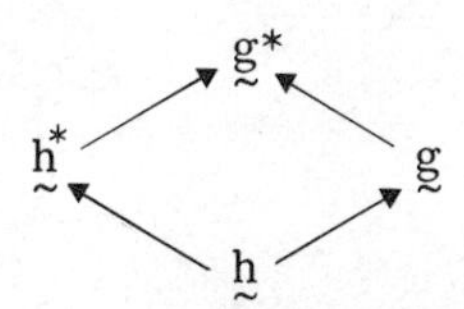

is commutative. Thus we can embed $\mathfrak{h}$, $\mathfrak{g}$, $\mathfrak{h}^*$ in $\mathfrak{g}^* = \mathfrak{g}^L$ without disturbing their interrelations in any way. The L-subspace of $\mathfrak{g}^L$ that is spanned by $\mathfrak{h}$ is a Lie algebra. <u>It is moreover the Lie algebra of</u> $H^{(L)}$.

6.5 A basic example

Throughout section (6.5) we shall be concerned with a non-trivial, unitary, and associative K-algebra A whose dimension (as a K-space) is finite. To avoid unimportant special cases it will be assumed, for the duration of section (6.5), that the ground field K is <u>infinite.</u>

It was shown, in section (5.3) Example 8, that the units of A form an affine group. As before this group will be denoted by U(A) and we supplement the notation by using $\underset{\sim}{u}(A)$ to describe the associated Lie algebra. If V is an n-dimensional $(n \geq 1)$ vector space over K, then we may take A to be $\mathrm{End}_K(V)$ in which case $U(A)$ becomes[†] $GL(V)$. In view of Theorem 27 of Chapter 5, it is clear that the study of the closed subgroups of $U(A)$ has important implications for the general theory.

Let $a \in A$. Then, as in (4.6.19), we can construct a derivation D_a of $K[A]$ over K. This extends naturally to a derivation of $K[U(A)]$ over K. It will be convenient to denote the extension by the same symbol. Thus $D_a \in \mathrm{Der}_K(K[U(A)])$ and if we put

$$A^* = \mathrm{Hom}_K(A, K), \tag{6.5.1}$$

then, by (4.6.20),

$$(D_a F)(b) = -F(ab) \tag{6.5.2}$$

for all $F \in A^*$ and $b \in A$. Again, by Chapter 4 Theorem 31,

$$(D_a P)(\sigma) = -(dP)(\sigma, a\sigma) \tag{6.5.3}$$

for all $P \in K[U(A)]$ and $\sigma \in U(A)$.

Lemma 3. _Let_ $\tau \in U(A)$. _Then_

$$(D_a P)^\tau = D_a(P^\tau) \tag{6.5.4}$$

for all $P \in K[U(A)]$.

Proof. If $F \in A^*$ and $\sigma \in U(A)$, then

$$((D_a F)^\tau)(\sigma) = (D_a F)(\sigma\tau) = -F(a\sigma\tau).$$

On the other hand $F^\tau \in A^*$ and therefore

$$(D_a F^\tau)(\sigma) = -F^\tau(a\sigma) = -F(a\sigma\tau).$$

---

† See the remarks following Theorem 19 of Chapter 5.

It follows that (6.5.4) holds for all $F \in A^*$ and it is clear that it holds for all constants. Again it is easy to see that if (6.5.4) holds for $P = P_1$ and $P = P_2$, then it also holds when $P = P_1 + P_2$ and when $P = P_1 P_2$. It follows that $(D_a P)^\tau = D_a(P^\tau)$ for all $P \in K[A]$ and now the extension to $K[U(A)]$ is immediate.

Lemma 3 shows that, for all $a \in A$, D_a belongs to the Lie algebra $\underset{\sim}{u}(A)$ of $U(A)$. We recall that A itself is a Lie algebra with $[a, b] = ab - ba$.

Theorem 21. The mapping

$$A \to \underset{\sim}{u}(A) \tag{6.5.5}$$

in which $a \mapsto D_a$ is an isomorphism of Lie algebras.

Proof. It is clear that (6.5.5) is K-linear. Also, by (6.5.2), if $D_a = 0$, then $F(a) = 0$ for all $F \in A^*$ and therefore $a = 0$. Consequently (6.5.5) is an injection. However A and $\underset{\sim}{u}(A)$ have the same dimension as K-spaces and therefore (6.5.5) is an isomorphism of K-spaces.

Let $a, c \in A$. The proof will be complete if we show that

$$D_{[a, c]} = [D_a, D_c]$$

and this will follow if we prove that

$$D_{[a, c]}F = [D_a, D_c]F$$

for all $F \in A^*$. Now, for $b \in A$,

$$\begin{aligned}(D_{[a, c]}F)(b) &= -F([a, c]b) \\ &= F(cab) - F(acb).\end{aligned}$$

On the other hand

$$([D_a, D_c]F)(b) = (D_a D_c F - D_c D_a F)(b).$$

But $D_c F \in A^*$. Consequently

$$(D_a D_c F)(b) = -(D_c F)(ab) = F(cab)$$

and likewise $(D_c D_a F)(b) = F(acb)$. The theorem follows.

On the basis of Theorem 21 we can identify $\underset{\sim}{u}(A)$ with A considered as a Lie algebra. Indeed in what follows we shall put

$$\underset{\sim}{u}(A) = A. \tag{6.5.6}$$

We next turn our attention to a typical closed subgroup G of U(A). As usual $\underset{\sim}{g}$ denotes the Lie algebra of G and in view of (6.5.6) _we may consider_ $\underset{\sim}{g}$ _as a Lie subalgebra of_ A. To simplify our notation a little we shall put

$$I(G) = I_{U(A)}(G). \tag{6.5.7}$$

Thus I(G) is an ideal of K[U(A)].

Theorem 22. _Let_ G _be a closed subgroup of_ U(A), _let_ $a \in A$ _and let_ $\sigma \in G$. _Then the following statements are equivalent_:

(1) $a \in \underset{\sim}{g}$;

(2) $(dP)(\sigma, a\sigma) = 0$ _for all_ $P \in I(G)$.

Proof. First suppose that $a \in \underset{\sim}{g}$ and $P \in I(G)$. By Theorem 11, $D_a P \in I(G)$ and therefore $(D_a P)(\sigma) = 0$. It follows, from (6.5.3), that $(dP)(\sigma, a\sigma) = 0$ and we have shown that (1) implies (2).

Next assume that (2) holds and let $P \in I(G)$. Then $P^{\sigma^{-1}} \in I(G)$ and therefore

$$(dP^{\sigma^{-1}})(\sigma, a\sigma) = 0,$$

that is to say $(D_a P^{\sigma^{-1}})(\sigma) = 0$. But

$$(D_a P^{\sigma^{-1}}) = (D_a P)^{\sigma^{-1}}$$

and thus $(D_a P)(1_A) = 0$. As this holds for all $P \in I(G)$ it follows, from Theorem 10, that $a \in \underset{\sim}{g}$.

There is a companion to Theorem 22 which will be proved shortly. First we recall that if $\sigma \in U(A)$, then _left_ translation by means of σ induces an automorphism λ^*_σ of the K-algebra K[U(A)]. Note that the inverse of λ^*_σ is $\lambda^*_{\sigma^{-1}}$ and that

$$\lambda^*_\sigma(A^*) = A^*. \tag{6.5.8}$$

Lemma 4. Let $a \in A$ and $\sigma \in U(A)$. Then

$$D_a \lambda^*_\sigma = \lambda^*_\sigma D_{\sigma a \sigma^{-1}}.$$

Proof. $D_{\sigma a \sigma^{-1}}$ and $\lambda^*_{\sigma^{-1}} D_a \lambda^*_\sigma$ both belong to $\mathrm{Der}_K(K[U(A)])$ and it will suffice to show that they are the same. For this it is sufficient to prove that they agree on A^*.

Assume therefore that $F \in A^*$ and $b \in A$. Then

$$\begin{aligned}((\lambda^*_{\sigma^{-1}} D_a \lambda^*_\sigma)F(b) &= (D_a(\lambda^*_\sigma F))(\sigma^{-1}b) \\ &= -(\lambda^*_\sigma F)(a\sigma^{-1}b)\end{aligned}$$

by (6.5.8) and (6.5.2). Furthermore

$$\begin{aligned}(\lambda^*_\sigma F)(a\sigma^{-1}b) &= F(\sigma a \sigma^{-1} b) \\ &= -(D_{\sigma a \sigma^{-1}} F)(b)\end{aligned}$$

and with this the lemma follows.

Corollary. If $P \in K[U(A)]$, then

$$(d\overline{\lambda^*_\sigma P})(1_A, a) = (dP)(\sigma, \sigma a). \tag{6.5.9}$$

Proof. We have

$$\begin{aligned}(d\overline{\lambda^*_\sigma P})(1_A, a) &= -((D_a \lambda^*_\sigma)P)(1_A) \\ &= -(\lambda^*_\sigma(D_{\sigma a \sigma^{-1}} P))(1_A) \\ &= -(D_{\sigma a \sigma^{-1}} P)(\sigma) \\ &= (dP)(\sigma, \sigma a)\end{aligned}$$

by (6.5.3).

The next result is the companion to Theorem 22.

Theorem 23. Let G be a closed subgroup of $U(A)$, let $a \in A$ and let $\sigma \in G$. Then the following statements are equivalent:

(1) $a \in \underset{\sim}{g}$;

(2) $(dP)(\sigma, \sigma a) = 0$ for all $P \in I(G)$.

Proof. Since $\sigma \in G$, it follows that $\lambda_\sigma^*(I(G)) = I(G)$. Consequently, by (6.5.9), condition (2) is equivalent to

$$(dP)(1_A, a) = 0$$

for all $P \in I(G)$. However, $1_A \in G$ and so the desired result follows from Theorem 22.

Now suppose that L is an extension field of K. Then A^L (see section (1.6)) is a non-trivial, unitary and associative L-algebra whose dimension as a vector space over L is finite. By Chapter 2 Theorem 32, we have $A^L = A^{(L)}$ because K is infinite, and an easy application of Lemma 14 of Chapter 2 shows that

$$U(A)^{(L)} = U(A^L).$$

Accordingly A^L is the Lie algebra of $U(A)^{(L)}$.

Suppose that $a \in A$. Then $a \in A^L$ so D_a can stand for either a derivation of $K[U(A)]$ over K or a derivation of $L[U(A^L)]$ over L. Now if $P \in K[U(A)]$ and $\hat{P}$ denotes its prolongation, then, as we saw at the end of section (4.6), $D_a\hat{P}$ is the prolongation of $D_a P$. This means that if D_a stands for the derivation of $K[U(A)]$ over K, then what we obtain by regarding a as belonging to A^L is D_a^L, where the notation is that used in Theorem 18.

Thus if G is a closed subgroup of $U(A)$, then $G^{(L)}$ is a closed subgroup of $U(A^L)$; and the Lie algebra $\underset{\sim}{g}^L$ of $G^{(L)}$ is obtained, with all its structure, by taking the L-subspace of A^L that is spanned by the K-subspace $\underset{\sim}{g}$ of A.

Now suppose that $D \in Der_K(L)$ and for $x \in A^L$ define $Dx \in A^L$ as in (4.6.13). We recall that if $x, x' \in A^L$ and $\lambda \in L$, then

$$D(x + x') = Dx + Dx',$$
$$D(\lambda x) = (D\lambda)x + \lambda(Dx),$$

and

$$D(b) = 0 \qquad (b \in A).$$

Every element of A^L can be written in the form $\lambda_1 b_1 + \lambda_2 b_2 + \ldots + \lambda_q b_q$ with $\lambda_i \in L$ and $b_i \in A$, and then

$$D(\lambda_1 b_1 + \lambda_2 b_2 + \ldots + \lambda_q b_q) = \sum_{i=1}^{q} (D\lambda_i) b_i \tag{6.5.10}$$

from which it follows that

$$D(xx') = (Dx)x' + x(Dx'). \tag{6.5.11}$$

Theorem 24. <u>Suppose that</u> $D \in \mathrm{Der}_K(L)$ <u>and that</u> $x \in G^{(L)}$. <u>Then</u> $(Dx)x^{-1}$ <u>belongs to</u> $\mathfrak{g}^L$.

Proof. Put $y = (Dx)x^{-1}$ and suppose that $P \in I(G)$. Then, with the usual notation for prolongations, $\hat{P}(x) = 0$ and therefore, by Chapter 4 Lemma 15, $(d\hat{P})(x, Dx) = 0$. Now the prolongations $\hat{P}$ generate the ideal $I(G)L[U(A^L)]$ and this is the ideal, $\mathfrak{U}$ say, of $L[U(A^L)]$ which is associated with $G^{(L)}$. Hence, by (4.6.8) and (4.6.9), we have $(dQ)(x, Dx) = 0$, that is $(dQ)(x, yx) = 0$, for all $Q \in \mathfrak{U}$. Consequently $y \in \mathfrak{g}^L$ by Theorem 22.

Theorem 25. <u>Let</u> G <u>be a closed connected subgroup of</u> $U(A)$, <u>and let</u> L <u>be an extension field of</u> K. <u>Further let</u> $x \in G^{(L)}$ <u>and be a generic point of</u> G. <u>Then given</u> $a \in \mathfrak{g}$ <u>there exists a unique</u> $D \in \mathrm{Der}_K(K(x))$ <u>such that</u> $Dx = ax$.

Proof. Since $K[G]$ is an integral domain, the derivation D_a extends to a derivation of $K(G)$ over K. (The extension will be denoted by the same symbol.) Next, because we have an isomorphism $K(G) \approx K(x)$ over K, D_a will induce a derivation, Δ say, of $K(x)$ over K. We put $D = -\Delta$.

Let $a_1, a_2, \ldots, a_n$ be a base for A over K and $F_1, F_2, \ldots, F_n$ the dual base. Then $a_1, a_2, \ldots, a_n$ is also a base for A^L over L and this time the prolongations $\hat{F}_1, \hat{F}_2, \ldots, \hat{F}_n$ form the dual base. Now

$$K(x) = K(\hat{F}_1(x), \hat{F}_2(x), \ldots, \hat{F}_n(x)) \tag{6.5.12}$$

and, for any $y \in A^L$,

$$y = \hat{F}_1(y)a_1 + \hat{F}_2(y)a_2 + \ldots + \hat{F}_n(y)a_n.$$

Next, by construction,

$$D(\hat{F}_i(x)) = -(D_a\hat{F}_i)(x) = \hat{F}_i(ax)$$

and therefore

$$\begin{aligned} Dx &= \sum_{i=1}^{n} (D\hat{F}_i(x))a_i \\ &= \sum_{i=1}^{n} \hat{F}_i(ax)a_i \\ &= ax. \end{aligned}$$

Finally if D' also belongs to $\mathrm{Der}_K(K(x))$ and $D'x = ax$, then $D(\hat{F}_i(x)) = D'(\hat{F}_i(x))$ for all i and therefore $D = D'$ by (6.5.12).

6.6 Further examples

In this section we shall give several examples of Lie algebras associated with affine groups. In many cases our account will depend on the results obtained in section (6.5).

Example 1. Let G be a finite group. Then $\mathrm{Dim}\, G = 0$ and therefore, by Theorem 13, its Lie algebra is zero-dimensional and hence null.

Example 2. Let V be an n-dimensional ($n \geq 1$) vector space over K. We know that $GL(V)$ is an affine group and we shall denote the associated Lie algebra by $\mathfrak{gl}(V)$. In considering this Lie algebra, Example 1 shows that we may assume that K is <u>infinite</u> for otherwise the Lie algebra is trivial.

Put $A = \mathrm{End}_K(V)$. Then A is a K-algebra to which the results of section (6.5) are applicable and we have already seen, in Example 8 of section (5.3), that

$$U(A) = GL(V). \qquad (6.6.1)$$

It therefore follows, from (6.5.6), that we may make the identification

$$\underset{\sim\sim}{gl}(V) = \mathrm{End}_K(V). \tag{6.6.2}$$

Of course if $f, g \in \mathrm{End}_K(V)$, then their Lie product is given by

$$[f, g] = f \circ g - g \circ f. \tag{6.6.3}$$

Example 3. Suppose that $n \geq 1$, let K be an <u>infinite</u> field, and denote the Lie group associated with the general linear group $GL_n(K)$ by $\underset{\sim\sim}{gl}_n(K)$. This situation is dealt with by taking A, in section (6.5), to be the K-algebra $M_n(K)$ of $n \times n$ matrices. If this is done, then $U(A) = GL_n(K)$ and therefore

$$\underset{\sim\sim}{gl}_n(K) = M_n(K). \tag{6.6.4}$$

Here if $A, B \in M_n(K)$, then their Lie product is

$$[A, B] = AB - BA. \tag{6.6.5}$$

Now let

$$X_{ij} : M_n(K) \to K \tag{6.6.6}$$

be the mapping which maps B, in $M_n(K)$, into its (i, j)-th entry. Then X_{ij} is a linear form on $M_n(K)$,

$$K[M_n(K)] = K(X_{11}, X_{12}, \ldots, X_{nn}],$$

and the X_{ij} are algebraically independent over K.

Suppose that $A \in M_n(K)$. By (6.5.2), the associated invariant derivation D_A, of $K[GL_n(K)]$, satisfies

$$(D_A X_{ij})(B) = -X_{ij}(AB), \tag{6.6.7}$$

for all B in $M_n(K)$.

We recall that the special linear group $SL_n(K)$ is a closed subgroup of $GL_n(K)$. Consequently its Lie algebra, which is denoted by $\underset{\sim\sim}{sl}_n(K)$, may be regarded as a Lie subalgebra of $M_n(K)$.

Lemma 5. Suppose that D_A, where A is the $n \times n$ matrix $\|a_{ij}\|$, belongs to $\underset{\sim\sim}{sl}_n(K)$. Then $\sum_{i=1}^{n} a_{ii} = 0$.

Proof. By Chapter 5 Theorem 17, the associated ideal, $\mathfrak{U}$ say, of $SL_n(K)$ in $K[GL_n(K)]$ is generated by $\mathrm{Det}\|X_{ij}\| - 1$. Consequently, by Theorem 11,

$$D_A(\mathrm{Det}\|X_{ij}\| - 1) \in \mathfrak{U}. \tag{6.6.8}$$

Put $Y_{ij} = X_{ij} - \delta_{ij}$. Then

$$\mathrm{Det}\|X_{ij}\| - 1 = Y_{11} + Y_{22} + \ldots + Y_{nn} + q(Y_{11}, Y_{12}, \ldots, Y_{nn}),$$

where $q(Y_{11}, \ldots, Y_{nn})$ belongs to $K[Y_{11}, Y_{12}, \ldots, Y_{nn}]$ and each of its terms has degree at least 2. Hence, if I denotes the $n \times n$ identity matrix, $Y_{ij}(I) = 0$ for all i, j and therefore

$$(D_A q(Y_{11}, Y_{12}, \ldots, Y_{nn}))(I) = 0.$$

Next, by (6.6.8),

$$(D_A(\mathrm{Det}\|X_{ij}\| - 1))(I) = 0$$

and so we see that

$$\sum_{i=1}^{n} (D_A Y_{ii})(I) = 0.$$

Finally, by (6.6.7),

$$(D_A Y_{ij})(I) = (D_A X_{ij})(I) = -a_{ij}$$

and therefore $a_{11} + a_{22} + \ldots + a_{nn} = 0$ as required.

Theorem 26. Let K be an infinite field and let the Lie algebra, $\underset{\sim\sim}{sl}_n(K)$, of $SL_n(K)$ be regarded as a Lie subalgebra of $M_n(K)$ in the manner explained above. Then $\underset{\sim\sim}{sl}_n(K)$ consists of all matrices with zero trace.

Proof. By Chapter 5 Theorem 17, $\mathrm{Dim}\, SL_n(K) = n^2 - 1$ and therefore $n^2 - 1$ is the dimension of $\underset{\sim\sim}{sl}_n(K)$ as a vector space over K.

Next Lemma 5 shows that if $A \in \underset{\sim\sim}{sl}_n(K)$, then its trace is equal to zero. The theorem now follows because the set of all matrices with zero trace forms a vector space of dimension $n^2 - 1$.

Example 4. In this example, K denotes an infinite field whose characteristic is different from 2. We suppose that $n \geq 1$ is an integer and that B is a non-singular $n \times n$ matrix with entries in K. In what follows we use the same notation as that previously employed in section (5.3) Example 9. In particular we put $G = GL_n(K)$ and

$$H = \{A \mid A \in M_n(K) \text{ and } A^T BA = B\}, \tag{6.6.9}$$

$$\Lambda = \{P \mid P \in M_n(K) \text{ and } P^T B + BP^T = 0\}. \tag{6.6.10}$$

(Here the superfix T denotes a transpose.) We already know that H is a closed subgroup of G and that by suitably choosing B we can obtain both the orthogonal and the symplectic groups. Our aim is to show that the Lie algebra $\underset{\sim}{h}$, of H, when considered as a Lie subalgebra of $\underset{\sim}{g} = M_n(K)$ is just Λ.

We recall that Λ is a vector space over K whose dimension is equal to Dim H (see Chapter 5 Theorem 20). If L is an extension field of K, then

$$M_n(K)^{(L)} = M_n(L),$$

and

$$G^{(L)} = GL_n(L).$$

Also in our discussion of H in Chapter 5 we showed how to choose L and $A^* \in M_n(L)$ so that A^* was a generic point for the connected component, H_0 say, of the identity of H.

Let C belong to $\underset{\sim}{h}$. Then C also belongs to the Lie algebra of H_0 and therefore, by Theorem 25, there exists $D \in \mathrm{Der}_K(K(A^*))$ such that

$$DA^* = CA^*. \tag{6.6.11}$$

Note that if $Q, Q' \in M_n(K(A^*))$ then, by (6.5.10), DQ is obtained by applying D to the entries in Q. Hence $D(Q^T) = (DQ)^T$,

$$D(QQ') = (DQ)Q' + Q(DQ')$$

and $DB = 0$. Now, by (5.3.9),

$$A^{*T}BA^* = B.$$

Consequently

$$(DA^*)^T BA^* + A^{*T}B(DA^*) = 0.$$

Thus

$$A^{*T}C^T BA^* + A^{*T}BCA^* = 0$$

whence

$$C^T B + BC = 0$$

because A^* and A^{*T} are invertible. Accordingly $C \in \Lambda$ and therefore $\underset{\sim}{h} \subseteq \Lambda$. However, by Chapter 5 Theorem 20, $\underset{\sim}{h}$ and Λ have the same dimension as K-spaces. Accordingly $\underset{\sim}{h} = \Lambda$ and we have proved

Theorem 27. *Let* K *be an infinite field whose characteristic is different from* 2, *let* $B \in GL_n(K)$ *and let* H *be the closed subgroup of* $GL_n(K)$ *defined by* (6.6.9). *If now the Lie algebra of* $GL_n(K)$ *is identified with* $M_n(K)$, *then the Lie algebra of its subgroup* H *is* Λ, *where* Λ *is defined in* (6.6.10).

Example 5. Let K be an *infinite* field and V an n-dimensional $(n \geq 1)$ vector space over K. Further let U and W be subspaces of V with $U \subseteq W$.

In Example 7 of section (5.3) we constructed a connected, closed subgroup $GL(U, W) = H$ (say) of $GL(V) = G$ (say). By (6.6.2), $g = \mathrm{End}_K(V)$. Our aim is to determine the Lie algebra of $GL(U, W)$ as a Lie subalgebra of $\mathrm{End}_K(V)$.

In what follows w denotes an element of W; $\xi \in \mathrm{Hom}_K(V, K)$ and satisfies $\xi(U) = 0$; and $\psi_{w,\xi}$ belongs to $\mathrm{Hom}_K(\mathrm{End}_K(V), K)$ and is defined by[†]

$$\psi_{w,\xi}(f) = \xi(fw).$$

Now when discussing Example 7 of section (5. 3) it was shown that the ideal $I_G(H)$ is generated by the functions $\psi_{w,\xi} - \xi(w)$. Let $f \in \mathrm{End}_K(V)$ and use I to denote the identity mapping of V. Then, by Theorem 10, $f \in \underset{\sim}{h}$ if and only if

$$(D_f(\psi_{w,\xi} - \xi(w))(I) = 0$$

for all w and ξ. Now, by (6. 5. 2) applied to the case where $A = \mathrm{End}_K(V)$,

$$\begin{aligned}(D_f(\psi_{w,\xi} - \xi(w))(I) &= (D_f\psi_{w,\xi})(I)\\ &= -\psi_{w,\xi}(f)\\ &= -\xi(fw).\end{aligned}$$

Accordingly $f \in \underset{\sim}{h}$ if and only if $fw \in U$ for all $w \in W$. This proves

Theorem 28. Suppose that K is an infinite field, V is an n-dimensional $(n \geq 1)$ vector space over K, and U, W are subspaces of V with $U \subseteq W$. Let $\underset{\sim}{gl}(V)$ be identified with $\mathrm{End}_K(V)$ and let $f \in \mathrm{End}_K(V)$. Then f belongs to the Lie algebra of $GL(U, W)$ if and only if $f(W) \subseteq U$.

6. 7 Adjoint representations

Let H be an affine group (defined over K), $\underset{\sim}{h}$ its Lie algebra, and let $x \in H$. The inner automorphism

$$\psi_x : H \to H \tag{6. 7. 1}$$

in which $y \mapsto xyx^{-1}$ induces an automorphism

$$d\psi_x : \underset{\sim}{h} \to \underset{\sim}{h} \tag{6. 7. 2}$$

---

† We use fw as an alternative to $f(w)$.

of the Lie algebra. Define

$$H \times \underset{\sim}{h} \to \underset{\sim}{h} \tag{6.7.3}$$

by $(x, a) \mapsto d\psi_x(a)$. Then $\underset{\sim}{h}$ becomes a left (H, K)-module and, in regar to this structure, we say that H acts on $\underset{\sim}{h}$ by conjugation. Arising fron this module structure we have a homomorphism

$$\mathrm{Ad} : H \to GL(\underset{\sim}{h}) \tag{6.7.4}$$

of abstract groups. This is a representation of H by means of automorphisms of $\underset{\sim}{h}$ and it is known as the adjoint representation of H. The main purpose of this section is to establish two claims concerning this representation. The first of these can be stated at once.

Claim I. The adjoint representation (6.7.4) is a rational represent tion.

Suppose, for the moment, that this has been established. Then from the adjoint representation of H we obtain a homomorphism

$$d(\mathrm{Ad}) : \underset{\sim}{h} \to \mathrm{End}_K(\underset{\sim}{h}) \tag{6.7.5}$$

of the Lie algebra of H into the Lie algebra of $GL(\underset{\sim}{h})$. In other words we obtain a representation of the Lie algebra $\underset{\sim}{h}$ by means of endomorphisms of $\underset{\sim}{h}$. We can now formulate our second claim.

Claim II. The representation of $\underset{\sim}{h}$ given by (6.7.5) is the same as the adjoint representation $\mathrm{ad} : \underset{\sim}{h} \to \mathrm{End}_K(\underset{\sim}{h})$ as defined in (6.2.4).

If K is a finite field, then both claims are trivial so let us assume that K is infinite. The aim of the remarks which follow is to show that we may suppose that K is algebraically closed.

Let L be an extension field of K and assume that $x \in H \subseteq H^{(L)}$. The L-homomorphism $H^{(L)} \to H^{(L)}$ in which $\eta \mapsto x\eta x^{-1}$ extends ψ_x and therefore it is just $\psi_x^{(L)}$. Moreover, by Theorem 20, $d\psi_x^{(L)} : \underset{\sim}{h}^L \to \underset{\sim}{h}^L$ extends $d\psi_x : \underset{\sim}{h} \to \underset{\sim}{h}$.

Let $e_1, e_2, \ldots, e_n$ be a base for $\underset{\sim}{h}$ over K. Then it is also a base for $\underset{\sim}{h}^L$ over L. Next

$$d\psi_x(e_j) = \sum_{i=1}^{n} \gamma_{ij}(x)e_i = d\psi_x^{(L)}(e_j),$$

where $\gamma_{ij} : H \to K$, and by Lemma 3 Chapter 5 and its corollary, the representation $Ad : H \to GL(\underset{\sim}{h})$ is rational if and only if $\gamma_{ij} \in K[H]$ for all i and j.

Assume that the adjoint representation $H^{(L)} \to GL(\underset{\sim}{h}^L)$ is rational and select i, j so that $1 \leq i, j \leq n$. Then there exists $F \in L[H^{(L)}]$ with the property that its restriction to H is γ_{ij}. Also we can find a base for L over K consisting of 1 and a family $\{\lambda_\alpha\}_{\alpha \in A}$ of elements of L. We can now write

$$F = \hat{f} + \sum_{\alpha \in A} \lambda_\alpha \hat{f}_\alpha.$$

(Here f and f_α belong to $K[H]$ and $\hat{f}$ and $\hat{f}_\alpha$ denote their natural prolongations to $H^{(L)}$. Also only finitely many of the f_α are non-zero.) This makes it clear that $\gamma_{ij} = f \in K[H]$ and the rationality of $Ad : H \to GL(\underset{\sim}{h})$ follows.

Next, because

$$GL(\underset{\sim}{h}^L) = (GL(\underset{\sim}{h}))^{(L)},$$

the adjoint representation $H^{(L)} \to GL(\underset{\sim}{h}^L)$ is none other than $(Ad)^{(L)}$ and, by Theorem 20,

$$d((Ad)^{(L)}) : \underset{\sim}{h}^L \to End_L(\underset{\sim}{h}^L)$$

extends $d(Ad) : \underset{\sim}{h} \to End_K(\underset{\sim}{h})$. Consequently if $d((Ad)^{(L)})$ is the adjoint representation of $\underset{\sim}{h}^L$, then $d(Ad)$ is the adjoint representation of $\underset{\sim}{h}$.

Let us sum up our conclusions so far. It has been shown that if our two claims hold for $H^{(L)}$ then they also hold for H. This in turn implies that, when we come to establish these claims, we may add the assumption that the ground field K is algebraically closed.

We now make a fresh start. It will be assumed that K is an <u>infinite</u>† field and that A denotes a K-algebra of the kind discussed in section (6. 5). Free use will be made of the results of that section and of the notation used in

---

† This assumption is to remain in force until we come to Theorem 29.

deriving them. In particular $U(A)$ will denote the affine group formed by the units of A and, as in (6.5.6), we shall identify its Lie algebra $\underset{\sim}{u}(A)$ with A itself. We now proceed to investigate the adjoint representation of $U(A)$.

Let $F \in A^*$, where $A^* = \mathrm{Hom}_K(A, K)$. Then

$$F(abc) = \sum_{i=1}^{s} F_i(a)F_i'(b)F_i''(c) \tag{6.7.6}$$

for all $a, b, c \in A$, where F_i, F_i', F_i'' belong to A^*. Next suppose that $x \in U(A)$, $a \in A$, and denote by D_a the corresponding invariant derivation of $K[U(A)]$. By (6.3.7), we have for the inner automorphism $\psi_x : U(A) \to U(A)$

$$\begin{aligned}((d\psi_x D_a)F)(1_A) &= (D_a(F \circ \psi_x))(1_A) \\ &= (D_a G)(1_A),\end{aligned}$$

where $G \in K[U(A)]$ and satisfies $G(\beta) = F(x\beta x^{-1})$. Accordingly

$$G = \sum_{i=1}^{s} F_i(x)F_i''(x^{-1})F_i'$$

and therefore, because $(D_a F_i')(1_A) = -F_i'(a)$,

$$\begin{aligned}((d\psi_x D_a)F)(1_A) &= -\sum_{i=1}^{s} F_i(x)F_i'(a)F_i''(x^{-1}) \\ &= -F(xax^{-1}) \\ &= (D_{xax^{-1}}F)(1_A).\end{aligned}$$

It follows that

$$((d\psi_x D_a)P)(1_A) = (D_{xax^{-1}}P)(1_A)$$

for all $P \in K[U(A)]$ whence $d\psi_x D_a = D_{xax^{-1}}$ by the corollary to Theorem 6. Consequently, when $\underset{\sim}{u}(A)$ is identified with A,

$$(d\psi_x)a = xax^{-1}. \tag{6.7.7}$$

Define

$$\omega : U(A) \times \underset{\sim}{u}(A) \to \underset{\sim}{u}(A)$$

by

$$\omega(x, a) = (d\psi_x)a = xax^{-1}.$$

Then

$$(F \circ \omega)(x, a) = F(xax^{-1}) = \sum_{i=1}^{s} F_i(x)F_i'(a)F_i''(x^{-1})$$

from which it is clear that $F \circ \omega \in K[U(A) \times \underset{\sim}{u}(A)]$. However A^* generates $K[A]$ as a K-algebra and so it follows that ω is a K-morphism of affine sets. Thus _when_ $U(A)$ _acts on_ $\underset{\sim}{u}(A)$ _by conjugation_ $\underset{\sim}{u}(A)$ _is a rational_ $U(A)$-_module._

We now turn our attention to the closed subgroups of $U(A)$.

Lemma 6. _Let_ H _be a closed subgroup of_ $U(A)$. _Then the adjoint representation_

$$Ad : H \to GL(\underset{\sim}{h}) \tag{6.7.8}$$

is a rational representation.

Proof. Let $x \in H$, define $\psi_x : H \to H$ as in (6.7.1), and $\hat{\psi}_x : U(A) \to U(A)$ by $\beta \mapsto x\beta x^{-1}$. Then $\hat{\psi}_x$ extends ψ_x and therefore $d\psi_x : \underset{\sim}{u}(A) \to \underset{\sim}{u}(A)$ extends $d\psi_x : \underset{\sim}{h} \to \underset{\sim}{h}$. Thus, by (6.7.7),

$$d\psi_x(a) = xax^{-1}$$

for all $a \in \underset{\sim}{h} \subseteq \underset{\sim}{u}(A) = A$. We know that when $U(A)$ acts on $\underset{\sim}{u}(A)$ by conjugation, this makes $\underset{\sim}{u}(A)$ a rational $U(A)$-module. It is now clear that the corresponding action of H on $\underset{\sim}{h}$ makes $\underset{\sim}{h}$ a rational H-module. However, by Chapter 5 Lemma 3 Cor., this is equivalent to our assertion.

Lemma 6 shows that we have a homomorphism $d(Ad)$ of $\underset{\sim}{h}$ into the Lie algebra $End_K(\underset{\sim}{h})$. The lemma which follows will help us to investigate $d(Ad)$.

Suppose that $F \in A^*$ and $a \in \underset{\sim}{h}$. Define

$$Q_{F,a} : End_K(\underset{\sim}{h}) \to K \tag{6.7.9}$$

by

$$Q_{F,a}(f) = F(f(a)). \tag{6.7.10}$$

Evidently $Q_{F,a}$ is a linear form on $\mathrm{End}_K(\underset{\sim}{h})$.

Lemma 7. <u>The</u> $Q_{F,a}$, <u>where</u> $F \in A^*$ <u>and</u> $a \in \underset{\sim}{h}$, <u>span the space of linear forms on</u> $\mathrm{End}_K(\underset{\sim}{h})$.

Proof. Let $a_1, a_2, \ldots, a_p$ be a base for $\underset{\sim}{h}$ and choose $F_1, F_2, \ldots, F_p$ in A^* so that their restrictions to $\underset{\sim}{h}$ form the dual base. Then the Q_{F_i, a_j} are linearly independent and the lemma follows.

To simplify the notation put $\phi = \mathrm{Ad}$ in (6.7.8) and note that for $a, b \in \underset{\sim}{h}$

$$((d\phi D_a)(Q_{F,b})(I) = (D_a(Q_{F,b} \circ \phi))(1_A),$$

where I denotes the identity mapping of $\underset{\sim}{h}$. Now when $x \in H$,

$$\begin{aligned} Q_{F,b}(\phi(x)) &= F((d\psi_x)b) \\ &= F(xbx^{-1}) \\ &= \sum_{i=1}^{s} F_i(x)F_i'(b)F_i''(x^{-1}), \end{aligned}$$

where we have reverted to the notation previously employed in (6.7.6). Hence, using a bar to indicate the restriction of a function to H,

$$Q_{F,b} \circ \phi = \sum_{i=1}^{s} F_i'(b)\overline{F}_i\overline{G}_i'',$$

where $G_i'' \in K[U(A)]$ and $G_i''(\beta) = F_i''(\beta^{-1})$. But

$$(D_a\overline{F}_i)(1_A) = (D_aF_i)(1_A) = -F_i(a)$$

and, by Theorem 14,

$$\begin{aligned} (D_a\overline{G}_i'')(1_A) &= (D_aG_i'')(1_A) \\ &= -(D_aF_i'')(1_A) \\ &= F_i''(a). \end{aligned}$$

Accordingly

$$((d\phi D_a)Q_{F,b})(I) = \sum_{i=1}^{s} F_i(1_A)F_i'(b)F_i''(a) - \sum_{i=1}^{s} F_i(a)F_i'(b)F_i''(1_A)$$
$$= F(ba) - F(ab).$$

Put $\lambda = ad(a)$. Then $\lambda(b) = ab - ba$ and, since $\lambda \in End_K(\underset{\sim}{h})$, it determines an invariant derivation, D_λ say, on $K[GL(\underset{\sim}{h})]$. Moreover

$$(D_\lambda Q_{F,b})(I) = -Q_{F,b}(\lambda) = F(ba - ab)$$

and now we see that

$$((d\phi D_a)Q_{F,b})(I) = (D_\lambda Q_{F,b})(I).$$

In view of Lemma 7 this immediately generalizes to

$$((d\phi D_a)P)(I) = (D_\lambda P)(I)$$

for all $P \in K[GL(\underset{\sim}{h})]$. However $d\phi D_a$ and D_λ are invariant derivations on $K[GL(\underset{\sim}{h})]$ and therefore

$$d\phi D_a = D_\lambda = D_{ad(a)}$$

by the corollary to Theorem 6. We are now ready to establish

Theorem 29. *Let H be an affine group over the (arbitrary) field K. Then*

$$Ad : H \to GL(\underset{\sim}{h})$$

is a rational representation of H and

$$d(Ad) : \underset{\sim}{h} \to End_K(\underset{\sim}{h})$$

coincides with $ad : \underset{\sim}{h} \to End_K(\underset{\sim}{h})$.

Proof. It was shown at the beginning of the section that it will suffice to prove the theorem when K is algebraically closed. But in that case Theorem 27 of Chapter 5 shows that H is isomorphic to a closed subgroup of U(A), where A is a suitable chosen K-algebra of the type we have just been considering. In view of this all the assertions follow from the previous discussion.

The next results give applications of the adjoint representations.

Theorem 30. _Suppose that the affine group_ G _is commutative. Then_ $[D, D'] = 0$ _for all_ $D, D' \in \underset{\sim}{g}$ _that is to say_ $\underset{\sim}{g}$ _is an abelian Lie algebra._

Proof. Since G is commutative, $Ad : G \to GL(\underset{\sim}{g})$ factors through a trivial group and therefore $d(Ad)$ is the null mapping of $\underset{\sim}{g}$ into $End_K(\underset{\sim}{g})$. The theorem follows, because, by Theorem 29, $d(Ad) = ad$.

We recall that if Λ is a Lie algebra over K, then an ideal of Λ is a subspace N such that $[n, x] \in N$ for all $n \in N$ and $x \in \Lambda$.

Theorem 31. _Let_ G _be an affine group and let_ H _be a closed normal subgroup. Then_ $\underset{\sim}{h}$ _is an ideal of the Lie algebra_ $\underset{\sim}{g}$.

Proof. We may suppose that the ground field is infinite. Let $x \in G$ and define the K-morphism $\psi_x : G \to G$ by $\psi_x(y) = xyx^{-1}$. Then $d\psi_x \in GL(\underset{\sim}{g})$ and $d\psi_x(\underset{\sim}{h}) \subseteq \underset{\sim}{h}$. Accordingly $d\psi_x \in GL(\underset{\sim}{h}, \underset{\sim}{h})$, where the notation is that employed in Example 7 of section (5.3). It follows that $Ad(G) \subseteq GL(\underset{\sim}{h}, \underset{\sim}{h})$ and therefore

$$ad : \underset{\sim}{g} \to End_K(\underset{\sim}{g})$$

satisfies $ad(\underset{\sim}{g}) \subseteq \underset{\sim}{gl}(\underset{\sim}{h}, \underset{\sim}{h})$. However Theorem 28 gives a full description of $\underset{\sim}{gl}(\underset{\sim}{h}, \underset{\sim}{h})$ and this shows that if $\alpha \in \underset{\sim}{g}$ and $\beta \in \underset{\sim}{h}$, then $[\alpha, \beta] \in \underset{\sim}{h}$. Consequently $\underset{\sim}{h}$ is an ideal of $\underset{\sim}{g}$.

7 · Power series and exponentials

General remarks

Let G be a connected affine group defined over a field K, and let $\underset{\sim}{g}$ be its Lie algebra. It is a fact that the connection between G and $\underset{\sim}{g}$ is particularly close when the characteristic of the ground field is zero, and in this chapter we shall develop a theory which explains why this is so. Some of our remarks apply regardless of the value of the characteristic of K. When we need to assume that the characteristic is zero this will be stated at the beginning of the relevant section, and the reader will also be reminded of this underlying assumption in the statement of theorems.

7.1 Rings of formal power series

Let K be a field and let $T_1, T_2, \ldots, T_q$ be indeterminates. The set of all power series in $T_1, T_2, \ldots, T_q$ will be denoted by $K[[T_1, T_2, \ldots, T_q]]$. Thus a typical member of $K[[T_1, T_2, \ldots, T_q]]$ is an infinite formal sum

$$\sum_{\nu_1 \geq 0, \ldots, \nu_q \geq 0} a_{\nu_1 \nu_2 \ldots \nu_q} T_1^{\nu_1} T_2^{\nu_2} \ldots T_q^{\nu_q},$$

where the ν_i are non-negative integers and the coefficients $a_{\nu_1 \nu_2 \ldots \nu_q}$ all belong to K. These power series can be added and multiplied in an obvious way and, as a result, $K[[T_1, T_2, \ldots, T_q]]$ becomes an integral domain which has K as a subfield.

Let us now put

$$S = K[[T_1, T_2, \ldots, T_q]]$$

and suppose that P belongs to S. Then P can be written (in a unique way) as an infinite formal sum

$$P = P_0 + P_1 + P_2 + \dots, \tag{7.1.1}$$

where $P_j = P_j(T_1, T_2, \dots, T_q)$ is a homogeneous polynomial of degree j in $T_1, T_2, \dots, T_q$ with coefficients in K. If P is not the null power series we put $|P| = \frac{1}{2^t}$, where t is the smallest non-negative integer such that $P_t \neq 0$, and if P is the null polynomial we put $|P| = 0$. We now have a real-valued function defined over S with the following properties:

$$|P| \leq 1 \ \underline{\text{for all}} \ P \in S; \tag{7.1.2}$$

$$|P| = 0 \ \underline{\text{when and only when}} \ P = 0; \tag{7.1.3}$$

$$|P \pm Q| \leq \max\{|P|, |Q|\} \ \underline{\text{for}} \ P, Q \in S; \tag{7.1.4}$$

$$|PQ| = |P||Q| \ \underline{\text{for}} \ P, Q \in S; \tag{7.1.5}$$

$$|a| = 1 \ \underline{\text{whenever}} \ a \in K, a \neq 0. \tag{7.1.6}$$

Note that (7.1.4) can be readily extended so that if $P, P', \dots, P^*$ all belong to S, then

$$|P \pm P' \pm \dots \pm P^*| \leq \max\{|P|, |P'|, \dots, |P^*|\}.$$

The next step is to turn S into a metric space by defining a distance function $d(P, Q)$ according to the formula

$$d(P, Q) = |P - Q|. \tag{7.1.7}$$

It is easily verified that this has the properties

$$d(P + P', Q + Q') \leq \max\{d(P, Q), d(P', Q')\} \tag{7.1.8}$$

and

$$d(PP', QQ') \leq \max\{d(P, Q), d(P', Q')\}, \tag{7.1.9}$$

from which it readily follows that <u>addition, subtraction and multiplication are continuous operations on</u> $K[[T_1, T_2, \dots, T_q]]$.

Now assume that we have an infinite sequence $\phi_0, \phi_1, \phi_2, \ldots$ of power series and that $d(\phi_\mu, \phi_\nu) \to 0$ as μ and ν tend to infinity independently. If $d(\phi_\mu, \phi_\nu) < \frac{1}{2^t}$, then the homogeneous constituents of ϕ_μ and ϕ_ν agree at least as far as the terms of degree t. For the moment let us keep t fixed. Then, for sufficiently large μ, the term of degree t in ϕ_μ does not depend on μ so we may denote it by P_t. Put

$$P = P_0 + P_1 + P_2 + \ldots$$

and note that the construction of P ensures that $d(\phi_\mu, P) \to 0$ as μ tends to infinity. This argument shows that, as a metric space, $K[[T_1, T_2, \ldots, T_q]]$ is complete.

The theory of convergence for formal power series is in some respects remarkably simple. This is exemplified by the following theorem.

Theorem 1. *Let $\phi_0, \phi_1, \phi_2, \ldots$ be an infinite sequence of formal power series in $T_1, T_2, \ldots, T_q$. Then the series*

$$\phi_0 + \phi_1 + \phi_2 + \phi_3 + \ldots$$

converges if and only if $\phi_h \to 0$ as $h \to \infty$. If the series does converge, then any rearrangement of it converges to the same sum.

Proof. It is clear that if the series converges, then $\phi_h \to 0$. Now assume that $\phi_h \to 0$. Put

$$\psi_m = \phi_0 + \phi_1 + \ldots + \phi_m.$$

Then for $m < n$ we have

$$d(\psi_m, \psi_n) = |\phi_{m+1} + \phi_{m+2} + \ldots + \phi_n|.$$

Hence, for $m \geq 0$ and $n \geq 0$, we have

$$d(\psi_m, \psi_n) \leq \sup\{|\phi_{k+1}|, |\phi_{k+2}|, \ldots\}, \tag{7.1.10}$$

where $k = \min(m, n)$, and the right hand side of (7.1.10) tends to zero as m and n tend to infinity. Since $K[[T_1, T_2, \ldots, T_q]]$ is complete the sequence $\psi_0, \psi_1, \psi_2, \ldots$ tends to a limit and therefore the original series converges.

Next suppose that the series

$$\phi_0 + \phi_1 + \phi_2 + \phi_3 + \ldots$$

converges and let

$$\phi'_0 + \phi'_1 + \phi'_2 + \phi'_3 + \ldots$$

be a rearrangement of it. If now $\varepsilon > 0$ is given we can choose k so that $|\phi_{k+1}|, |\phi_{k+2}|, |\phi_{k+3}|, \ldots$ are all smaller than ε. This done we choose μ_0 so that $\phi_0, \phi_1, \phi_2, \ldots, \phi_k$ occur among $\phi'_0, \phi'_1, \ldots, \phi'_\mu$ whenever $\mu > \mu_0$. Then, provided that $\mu > \mu_0$,

$$|(\phi_0 + \phi_1 + \ldots + \phi_\mu) - (\phi'_0 + \phi'_1 + \ldots + \phi'_\mu)| \leq \sup\{|\phi_{k+1}|, |\phi_{k+2}|, \ldots\} <$$

This shows that

$$(\phi_0 + \phi_1 + \ldots + \phi_\mu) - (\phi'_0 + \phi'_1 + \ldots + \phi'_\mu)$$

tends to zero as $\mu \to \infty$ and from this we conclude that the rearranged series not only converges but that it has the same sum as the series with which we started.

At this point it is convenient to insert a few remarks about power series in a single indeterminate. The indeterminate will be denoted by T so, for the moment, we are concerned with the ring $K[[T]]$.

Suppose that the power series

$$\phi = a_0 + a_1 T + a_2 T^2 + a_3 T^3 + \ldots$$

belongs to $K[[T]]$. If ϕ is a unit in the ring then clearly $a_0 \neq 0$. Now suppose that $a_0 \neq 0$. If

$$\psi = b_0 + b_1 T + b_2 T^2 + b_3 T^3 + \ldots$$

then $\phi\psi = 1$ provided that the equations

$$a_0 b_0 = 1,$$

$$a_0 b_1 + a_1 b_0 = 0,$$

$$a_0 b_2 + a_1 b_1 + a_2 b_0 = 0,$$

$$\cdots\cdots\cdots\cdots$$

all hold. But these equations can be solved in succession to yield $b_0, b_1, b_2, \ldots$ and in this way an inverse for ϕ is obtained. Thus, to sum up, ϕ *is a unit in* $K[[T]]$ *if and only if* $a_0 \neq 0$.

Now assume that $\phi \in K[[T]]$ and $\phi \neq 0$. Then we can write

$$\phi = a_r T^r + a_{r+1} T^{r+1} + a_{r+2} T^{r+2} + \ldots,$$

where $a_r \neq 0$. Thus $\phi = T^r(a_r + a_{r+t} T + a_{r+2} T^2 + \ldots)$ and therefore $\phi = T^r \eta$ where η is a unit in $K[[T]]$.

Theorem 2. *Let* $\mathfrak{U}$ *be a non-zero ideal of* $K[[T]]$. *Then* $\mathfrak{U} = (T^s)$ *for a unique* $s \geq 0$.

Proof. The remarks just preceding the statement of the theorem show that $\mathfrak{U}$ contains a power of T. Let T^s be the first power of T to belong to $\mathfrak{U}$. Now suppose that $\phi \in \mathfrak{U}$ and $\phi \neq 0$. Then $\phi = T^r \eta$ for some $r \geq 0$ and some unit η of $K[[T]]$. It follows that $T^r \in \mathfrak{U}$ and therefore $r \geq s$. Thus $\phi \in (T^s)$ and now we see that $\mathfrak{U} = (T^s)$. The assertion concerning uniqueness is clear.

7.2 Modules over a power series ring

Once again let $T_1, T_2, \ldots, T_q$ be indeterminates and put

$$S = K[[T_1, T_2, \ldots, T_q]].$$

Since S is an integral domain it has a quotient field L say, and this is an extension field of K.

Now suppose that V is a finite-dimensional vector space over K and define the L-space V^L as in section (1.6). Then V^L can be regarded as an S-module in which V is embedded. The S-submodule of V^L that is generated by V will be denoted by V^S.

Put $V^* = \mathrm{Hom}_K(V, K)$ and let $F \in V^*$. We know from section (1.6) that F has a unique extension to an L-linear mapping of V^L into $K^L = L$. (The extension will also be denoted by F.) Choose a base $v_1, v_2, \ldots, v_n$ for V over K. Then each element of V^S has a unique representation in the form $\alpha_1 v_1 + \alpha_2 v_2 + \ldots + \alpha_n v_n$, where $\alpha_i \in S$. Since

$$F(\alpha_1 v_1 + \ldots + \alpha_n v_n) = \alpha_1 F(v_1) + \alpha_2 F(v_2) + \ldots + \alpha_n F(v_n),$$

it follows that $F(V^S) \subseteq S$.

For $x \in V^S$ put

$$|x| = \sup_{F \in V^*} |F(x)| \tag{7.2.1}$$

and observe that, since $|F(x)|$ is either zero or $\frac{1}{2^t}$ for some $t \geq 0$, the supremum is attained. Observe too that

$$|x| \leq 1. \tag{7.2.2}$$

Lemma 1. _Let_ $v_1, v_2, \ldots, v_n$ _be a base for_ V _over_ K _and let_ $\alpha_1, \alpha_2, \ldots, \alpha_n$ _belong to_ S. _Then_

$$|\alpha_1 v_1 + \alpha_2 v_2 + \ldots + \alpha_n v_n| = \max\{|\alpha_1|, |\alpha_2|, \ldots, |\alpha_n|\}.$$

Proof. Let $F \in V^*$. Then

$$\begin{aligned} &|F(\alpha_1 v_1 + \alpha_2 v_2 + \ldots + \alpha_n v_n)| \\ &\quad = |\alpha_1 F(v_1) + \alpha_2 F(v_2) + \ldots + \alpha_n F(v_n)| \\ &\quad \leq \max\{|\alpha_1||F(v_1)|, |\alpha_2||F(v_2)|, \ldots, |\alpha_n||F(v_n)|\} \\ &\quad \leq \max\{|\alpha_1|, |\alpha_2|, \ldots, |\alpha_n|\} \end{aligned}$$

because $|F(v_i)| \leq 1$. Now let $F_1, F_2, \ldots, F_n$ be the K-base of V^* that is dual to $v_1, v_2, \ldots, v_n$. Then

$$\begin{aligned} |\alpha_1 v_1 + \alpha_2 v_2 + \ldots + \alpha_n v_n| &\geq |F_i(\alpha_1 v_1 + \alpha_2 v_2 + \ldots + \alpha_n v_n)| \\ &= |\alpha_i| \end{aligned}$$

and now the lemma follows.

Lemma 2. Let $x, y \in V^S$ and let $\alpha, \beta \in S$. Then

(a) $|x| \leq 1$,

(b) $|x \pm y| \leq \max\{|x|, |y|\}$,

(c) $|\alpha x| = |\alpha||x|$,

(d) $|\alpha x - \beta y| \leq \max\{|x-y|, |\alpha-\beta|\}$.

Proof. The assertion (a) has already been established and (b) and (c) follow from Lemma 1. Finally

$$\begin{aligned}|\alpha x - \beta y| &= |\alpha(x - y) + (\alpha - \beta)y| \\ &\leq \max\{|\alpha||x - y|, |\alpha - \beta||y|\} \\ &\leq \max\{|x - y|, |\alpha - \beta|\}.\end{aligned}$$

In addition to the formulae contained in Lemma 2, we note that we can add

$$|x| = 1 \text{ whenever } x \in V \text{ and } x \neq 0. \tag{7.2.3}$$

The S-module V^S becomes a complete metric space if we define the distance between x and y to be $|x - y|$. Addition and subtraction are now continuous operations on V^S. Also if $x_n \to x$ in V^S and $\alpha_n \to \alpha$ in S, then $\alpha_n x_n \to \alpha x$.

Theorem 3. With the above notation let $x_0, x_1, x_2, \ldots,$ be an infinite sequence of elements of V^S. Then the series

$$x_0 + x_1 + x_2 + x_3 + \ldots$$

converges if and only if $x_n \to 0$ as $n \to \infty$. Should the series converge, then any rearrangement of it converges to the same sum.

The proof is virtually the same as that of Theorem 1 so we shall not give any details.

As before let V be a finite-dimensional vector space over K but now suppose that $D \in \mathrm{Der}_K(L)$. Then for $x \in V^L$ we can define $Dx \in V^L$ by using the construction first described (in a slightly different context) in (4.6.13). This leads to a mapping $D : V^L \to V^L$ with the following properties. For $x, x' \in V^L$, $\lambda \in L$, $k \in K$ and $v \in V$:

$$D(x + x') = Dx + Dx', \tag{7.2.4}$$

$$D(\lambda x) = (D\lambda)x + \lambda(Dx), \tag{7.2.5}$$

$$D(kx) = k(Dx), \tag{7.2.6}$$

$$D(v) = 0. \tag{7.2.7}$$

Hence if $u_1, u_2, \ldots, u_s$ belong to V and $\eta_1, \eta_2, \ldots, \eta_s$ belong to L then

$$D(\eta_1 u_1 + \eta_2 u_2 + \ldots + \eta_s u_s) = (D\eta_1)u_1 + (D\eta_2)u_2 + \ldots + (D\eta_s)u_s. \tag{7.2.8}$$

Note that if the derivation D satisfies $D(S) \subseteq S$, then $D(V^S) \subseteq V^S$. Moreover if the mapping $S \to S$ induced by D is continuous, then the mapping $V^S \to V^S$ in which $x \mapsto Dx$ is continuous as well.

We must now give some attention to the case where V is not merely a finite-dimensional K-space but has the further property of being a general K-algebra. In this situation we know, from section (6.1), that V^L is a general L-algebra. Clearly if $x, y \in V^S$ then the product $xy \in V^S$.

Suppose next that $D \in \mathrm{Der}_K(L)$ and $D(S) \subseteq S$. Then we have

$$D(xy) = (Dx)y + x(Dy) \tag{7.2.9}$$

whenever x and y belong to V^S. (This may be seen by expressing x and y as linear combinations of elements of V (with coefficients in S) and applying the formula (7.2.8).)

Finally suppose that $v_1, v_2, \ldots, v_n$ is a base for V over K and let

$$x = \alpha_1 v_1 + \alpha_2 v_2 + \ldots + \alpha_n v_n,$$
$$y = \beta_1 v_1 + \beta_2 v_2 + \ldots + \beta_n v_n,$$

where α_i and β_i belong to S. We know, from Lemma 1, that

$$|x| = \max\{|\alpha_1|, |\alpha_2|, \ldots, |\alpha_n|\}$$

and

$$|y| = \max\{|\beta_1|, |\beta_2|, \ldots, |\beta_n|\}.$$

But

$$xy = \sum_i \sum_j \alpha_i\beta_j v_i v_j,$$

and $|v_i v_j| \le 1$. Thus $|xy|$ is at most equal to the maximum of the numbers $|\alpha_i\beta_j|$ and therefore

$$|xy| \le |x||y| \qquad (7.2.10)$$

for all x, y in V^S. It follows that

$$|xy - x'y'| \le \max\{|x - x'|, |y - y'|\} \qquad (7.2.11)$$

whenever x, y, x', y' are in V^S and therefore multiplication is continuous on V^S.

7.3 Exponentials

Throughout section (7.3) we shall assume that the characteristic of K is zero and A will be used consistently to denote a K-algebra which is non-trivial, unitary and associative (but not necessarily commutative) and whose dimension as a K-space is finite. As before we set

$$S = K[[T_1, T_2, \ldots, T_q]],$$

where $T_1, T_2, \ldots, T_q$ are indeterminates, and we use L to denote the quotient field of S.

Algebras such as A have already received considerable attention in these pages. In Example 8 of section (5.3) we introduced a function $N : A \to K$ which is useful in the study of the units of A. Let us call this the norm function on A and recall that (i) $N \in K[A]$, and (ii) an element α (of A) is a unit of A if and only if $N(\alpha) \ne 0$. Of course the L-algebra A^L has its own norm function. However this coincides with N on A so we may also use N to denote the norm function of the larger algebra.

Next, because A^S is closed under multiplication, it follows that $N(A^S) \subseteq S$. But N is a multiplicative function and $N(1_A) = 1_K$. Consequently if x is a unit of A^S, then $N(x)$ is a unit of S.

After these preliminaries suppose that $x \in A^S$ and $|x| < 1$. By (7.2.10), we have $|x^\nu| \leq |x|^\nu$ for $\nu \geq 0$ and therefore the series $\sum \frac{x^\nu}{\nu!}$ converges. Put

$$\exp x = \sum_{\nu=0}^{\infty} \frac{x^\nu}{\nu!} \qquad (|x| < 1). \tag{7.3.1}$$

Then $\exp x \in A^S$.

Next let x and y be elements of A^S such that $|x| < 1$ and $|y| < 1$. By Lemma 2, we also have $|x + y| < 1$ and therefore $\exp x$, $\exp y$ and $\exp(x + y)$ are all defined. Indeed if x and y commute, then

$$\exp(x + y) = (\exp x)(\exp y) \tag{7.3.2}$$

by virtue of considerations very similar to those encountered in the classical theory of the exponential function. Since x and $-x$ commute and $\exp(0) = 1$, it follows that

$$(\exp x)(\exp(-x)) = 1 = (\exp(-x))(\exp x). \tag{7.3.3}$$

Thus, when $|x| < 1$, $\exp x$ is a unit of A^S and $N(\exp x)$ is a unit in S.

The power series ring S is a K-algebra. Let $D \in \mathrm{Der}_K(S)$ and observe that D has a natural extension to a derivation of L over K. (We denote this extension by the same letter.) For $x \in A^L$ we define Dx as in section (7.2). Of course $D(A^S) \subseteq A^S$. As we noted earlier, if the derivation $D : S \to S$ is continuous, then the induced mapping $A^S \to A^S$ is continuous as well.

Theorem 4. Let $D \in \mathrm{Der}_K(S)$ and be continuous. If now $x \in A^S$, $|x| < 1$ and x and Dx commute, then

$$D(\exp x) = (\exp x)(Dx) = (Dx)(\exp x).$$

Proof. Since x and Dx commute, we have, by virtue of (7.2.9),

$$D\left(\sum_{\nu=0}^{m} \frac{x^{\nu}}{\nu!}\right) = \left(\sum_{\nu=0}^{m-1} \frac{x^{\nu-1}}{(\nu-1)!}\right)(Dx) = (Dx)\left(\sum_{\nu=0}^{m-1} \frac{x^{\nu-1}}{(\nu-1)!}\right)$$

for all $m \geq 1$. But multiplication on A^S is continuous and so too is the mapping $A^S \to A^S$ induced by D. The theorem therefore follows by letting m tend to infinity.

Now put

$$D_i = \frac{\partial}{\partial T_i} \tag{7.3.4}$$

for $i = 1, 2, \ldots, q$. We may regard D_i as belonging either to $\mathrm{Der}_K(S)$ or to $\mathrm{Der}_K(L)$. As a member of $\mathrm{Der}_K(S)$, the derivation D_i is continuous. Of course $D_iT_i = 1$ whereas $D_iT_j = 0$ when $i \neq j$.

Suppose that $a_1, a_2, \ldots, a_q$ belong to A. Then $|a_iT_i| < 1$ and therefore $\exp(a_iT_i)$ is defined. Moreover a_iT_i and $D_j(a_iT_i)$ commute in A^S. Consequently, by Theorem 4,

$$D_i(\exp(a_iT_i)) = a_i(\exp(a_iT_i)) = (\exp(a_iT_i))a_i \tag{7.3.5}$$

and

$$D_j(\exp(a_iT_i)) = 0 \ \underline{\text{for}} \ i \neq j. \tag{7.3.6}$$

Moreover $\exp(a_iT_i)$ is a unit of A^S.

We now put

$$\omega = \exp(a_1T_1)\exp(a_2T_2) \ldots \exp(a_qT_q)$$

and, for $1 \leq i \leq q$,

$$\omega_i = \exp(a_1T_1)\exp(a_2T_2) \ldots \exp(a_iT_i),$$
$$\omega_i' = \exp(a_{i+1}T_{i+1})\exp(a_{i+2}T_{i+2}) \ldots \exp(a_qT_q),$$

where by ω_q' is to be understood 1_A. Then $\omega = \omega_i\omega_i'$ and therefore, by (7.2.9), (7.3.5) and (7.3.6),

$$\begin{aligned} D_i\omega &= (D_i\omega_i)\omega_i' + \omega_i(D_i\omega_i') \\ &= (D_i\omega_i)\omega_i' \\ &= \omega_i a_i \omega_i'. \end{aligned}$$

It follows that

$$(D_i \omega)\omega^{-1} = \omega_i a_i \omega_i^{-1}.$$

Of course ω, ω^{-1}, ω_i, ω_i^{-1} and a_i all belong to A^S.

Lemma 3. <u>Assume that</u> $a_1, a_2, \ldots, a_q$ <u>belong to</u> A <u>and are linearly independent over</u> K. <u>Then (with the above notation) the elements</u>

$$(D_i \omega)\omega^{-1} = \omega_i a_i \omega_i^{-1} \quad (i = 1, 2, \ldots, q)$$

<u>are linearly independent over</u> L.

Proof. Let $\phi_1, \phi_2, \ldots, \phi_q$ belong to S. We shall prove that

$$\left| \sum_{i=1}^{q} \phi_i \omega_i a_i \omega_i^{-1} \right| = \max \left\{ |\phi_1|, |\phi_2|, \ldots, |\phi_q| \right\} \tag{7.3.7}$$

and from this the lemma will follow.

It is clear that $|\exp(a_j T_j) - 1_A| < 1$ and since $(\exp(a_j T_j))^{-1} = \exp(-a_j T_j)$ we have $|(\exp(a_j T_j))^{-1} - 1_A| < 1$ as well. Next successive applications of (7.2.11) show that, for $\xi_1, \xi_2, \ldots, \xi_p$ and $\eta_1, \eta_2, \ldots, \eta_p$ in A^S, we have

$$|\xi_1 \xi_2 \ldots \xi_p - \eta_1 \eta_2 \ldots \eta_p| \leq \max\{|\xi_1 - \eta_1|, |\xi_2 - \eta_2|, \ldots, |\xi_p - \eta_p|\}$$

so we now see that $|\omega_i - 1_A| < 1$ and $|\omega_i^{-1} - 1_A| < 1$. From this it follows that

$$|\omega_i a_i \omega_i^{-1} - a_i| < 1. \tag{7.3.8}$$

Since our aim is to establish (7.3.7), we may suppose that at least one ϕ_i is not zero. Put

$$\max\{|\phi_1|, |\phi_2|, \ldots, |\phi_q|\} = \frac{1}{2^m}.$$

Then $|\phi_i(\omega_i a_i \omega_i^{-1} - a_i)| < \frac{1}{2^m}$ and therefore

$$\left| \sum_{i=1}^{q} \phi_i (\omega_i a_i \omega_i^{-1} - a_i) \right| < \frac{1}{2^m}. \tag{7.3.9}$$

On the other hand $a_1, a_2, \ldots, a_q$ are linearly independent over K and therefore

$$\left|\sum_{i=1}^{q} \phi_i a_i\right| = \max\{|\phi_1|, |\phi_2|, \ldots, |\phi_q|\} = \frac{1}{2^m}.$$

It follows from this and (7. 3. 9) that

$$\left|\sum_{i=1}^{q} \phi_i \omega_i a_i \omega_i^{-1}\right| = \frac{1}{2^m}$$

and with this the proof is complete.

We are now ready to prove

Theorem 5. Suppose that K has characteristic zero and that A is a non-trivial, unitary and associative K-algebra. If now $a_1, a_2, \ldots, a_q$ are elements of A that are linearly dependent over K, and

$$\omega = \exp(a_1 T_1)\exp(a_2 T_2)\ldots\exp(a_q T_q),$$

then ω is a unit of A^S (and hence of A^L) and the transcendence degree of $K(\omega)$ over K is at least q. (Here L is the quotient field of the ring $S = K[[T_1, T_2, \ldots, T_q]]$ of formal power series in $T_1, T_2, \ldots, T_q$.)

Proof. Since the characteristic of K is zero it follows, from Chapter 4 Theorem 12, that the transcendence degree of $K(\omega)$ over K is equal to the dimension of $\mathrm{Der}_K(K(\omega), L)$ considered as an L-space. As before put $D_i = \partial/\partial T_i$ and let Δ_i be the restriction of D_i to $K(\omega)$. Then $\Delta_1, \Delta_2, \ldots, \Delta_q$ belong to $\mathrm{Der}_K(K(\omega), L)$ and it will suffice to show that they are linearly independent with respect to L.

Suppose therefore that $\lambda_1\Delta_1 + \lambda_2\Delta_2 + \ldots + \lambda_q\Delta_q = 0$, where $\lambda_i \in L$. Since $\omega \in A^S$ we can write it in the form

$$\omega = \phi_1 b_1 + \phi_2 b_2 + \ldots + \phi_q b_q,$$

where $\phi_1, \phi_2, \ldots, \phi_q$ belong to S and $b_1, b_2, \ldots, b_q$ both belong to A and are linearly independent over K.

Now let $F_1, F_2, \ldots, F_q$ be K-linear mappings of A into K such that $F_i(b_i) = 1$ and $F_i(b_j) = 0$ whenever $i \neq j$. Then $F_i \in K[A]$ and has a natural prolongation $\hat{F}_i$ to a member of $L[A^L] = L[A^{(L)}]$. Also

$\hat{F}_i(\omega) = \phi_i$ and therefore $\phi_i \in K(\omega)$. Accordingly

$$\lambda_1(D_1\phi_i) + \lambda_2(D_2\phi_i) + \ldots + \lambda_q(D_q\phi_i) = 0$$

whence

$$\lambda_1(D_1\omega) + \lambda_2(D_2\omega) + \ldots + \lambda_q(D_q\omega) = 0$$

because

$$D_j\omega = (D_j\phi_1)b_1 + (D_j\phi_2)b_2 + \ldots + (D_j\phi_q)b_q$$

for all j. Thus

$$\lambda_1((D_1\omega)\omega^{-1}) + \lambda_2((D_2\omega)\omega^{-1}) + \ldots + \lambda_q((D_q\omega)\omega^{-1}) = 0$$

and therefore, by Lemma 3, the λ_i are all zero. The theorem follows.

Let U(A) be the group of units of A. This group has already received considerable attention particularly in section (6. 5). We recall some basic facts.

The group can be regarded as an affine group with coordinate ring K[A][1/N], where N denotes the norm function, and $(U(A))^{(\Omega)} = U(A^{\Omega})$ for any extension field Ω of K. Also the Lie algebra $\underset{\sim}{u}(A)$, of U(A), may be identified with A (considered as a Lie algebra). Consequently if H is a closed subgroup of U(A), then it is possible to regard its Lie algebra $\underset{\sim}{h}$ as a subalgebra of A.

Theorem 6. Let H be a closed subgroup of U(A) and let $a_1, a_2, \ldots, a_q$ belong to the Lie algebra $\underset{\sim}{h}$ of H. Put

$$\omega = \exp(a_1T_1)\exp(a_2T_2) \ldots \exp(a_qT_q).$$

Then $\omega \in H^{(L)}$ and therefore ω is a generalized point of H.

Proof. Because $H^{(L)}$ is a group it is enough to show that $\exp(a_1T_1)$ belongs to $H^{(L)}$. Put $S_1 = K[[T_1]]$ and let L_1 be its quotient field. Then $\exp(a_1T_1)$ is a unit of A^{L_1} and it will suffice to show that it belongs to $H^{(L_1)}$. Hence for the remainder of the proof we may suppose that $q = 1$, i.e. that we are dealing with just one indeterminate. This indeterminate will be denoted by T and we shall write a in place of a_1.

Let $Q \in K[U(A)]$. Then $Q = P/N^h$, where $P \in K[A]$, $h > 0$ and N is the norm function. Now Q and P have natural prolongations, $\hat{Q}$ and $\hat{P}$ say, to coordinate functions on $(U(A))^{(L)} = U(A^L)$. It is clear that $\hat{P}(\exp(aT)) \in S$ and we have noted earlier that $N(\exp(aT))$ is a unit of S. Consequently $\hat{Q}(\exp(aT)) \in S$.

Let $\mathfrak{U}$ be the ideal of H in $K[U(A)]$ and denote by $\mathfrak{U}'$ the ideal of S that is generated by the elements $\hat{Q}(\exp(aT))$ as Q varies over $\mathfrak{U}$. Put $D = \frac{\partial}{\partial T}$ and denote by D_a the invariant derivation, of $K[U(A)]$ or $L[(U(A))^{(L)}]$, that is associated with the element a. (Note that $a \in A \subseteq A^L$.) By Chapter 4 Lemma 15,

$$\begin{aligned} D(\hat{Q}(\exp(aT))) &= (d\hat{Q})(\exp(aT), D\exp(aT)) \\ &= (d\hat{Q})(\exp(aT), a\exp(aT)) \\ &= -(D_a\hat{Q})(\exp(aT)). \end{aligned}$$

(Here we have made use of Theorem 4 of this chapter and Theorem 31 of Chapter 4.) But, as we saw in section (4.6), $D_a\hat{Q}$ is the natural prolongation of D_aQ and, since $a \in \underset{\sim}{h}$, we have $D_a(\mathfrak{U}) \subseteq \mathfrak{U}$ by Chapter 6 Theorem 11. Accordingly for $Q \in \mathfrak{U}$

$$(D_a\hat{Q})(\exp(aT)) \in \mathfrak{U}'$$

and now it follows that $D(\mathfrak{U}') \subseteq \mathfrak{U}'$.

<u>We claim that</u> $\mathfrak{U}' = (0)$. For assume the contrary. Then, by Theorem 2, $\mathfrak{U}' = T^m K[[T]]$ for some $m \geq 0$. Now $\exp(aT) - 1_A$ is in TA^S so if (with our earlier notation) $Q = P/N^h$ belongs to $\mathfrak{U}$, then $P(1_A) = 0$ and therefore

$$\hat{P}(\exp(aT)) = \hat{P}(\exp(aT) - 1_A + 1_A)$$

belongs to $TK[[T]]$. Accordingly

$$\hat{Q}(\exp(aT)) \in TK[[T]]$$

and this shows that $m \geq 1$. Next $T^m \in \mathfrak{U}'$. It follows that $D(T^m) \in \mathfrak{U}'$ that is to say $mT^{m-1} \in \mathfrak{U}'$. But K has characteristic zero so $T^{m-1} \in \mathfrak{U}'$. This gives a contradiction and thereby establishes our claim that $\mathfrak{U}' = (0)$.

It has now been established that $\hat{Q}(\exp(aT)) = 0$ whenever $Q \in \mathfrak{U}$. Accordingly $\exp(aT) \in H^{(L)}$ by Chapter 2 Theorem 34.

Theorem 7. _Let K have characteristic zero and let H be a closed connected subgroup of U(A). Let $\underset{\sim}{h} \subseteq A$ be the Lie algebra of H and let $a_1, a_2, \ldots, a_q$ be a base for $\underset{\sim}{h}$ over K. Furthermore suppose that $T_1, T_2, \ldots, T_q$ are indeterminates and denote by L the quotient field of $K[[T_1, T_2, \ldots, T_q]]$. Then_

$$\omega = \exp(a_1 T_1)\exp(a_2 T_2) \ldots \exp(a_q T_q)$$

belongs to $H^{(L)}$ and is a generic point of H.

Proof. Since q is the dimension of $\underset{\sim}{h}$ as a K-space, we have $\text{Dim } H = q$ and we also have $\omega \in H^{(L)}$ by Theorem 6. Next the transcendence degree of $K(\omega)$ over K is at least q by Theorem 5, whereas Theorem 23 of Chapter 3 shows that it does not exceed this value. In fact the result just quoted shows that ω is a generic point of H as we wished to prove.

The lemma which follows will be superseded by an important general result to be found in the next section. The role of the lemma is to provide a stepping-stone to this result.

Lemma 4. _Let H_1 and H_2 be closed connected subgroups of U(A) and let $\underset{\sim}{h}_1$ and $\underset{\sim}{h}_2$ be their Lie algebras. If now $\underset{\sim}{h}_1 \subseteq \underset{\sim}{h}_2$, then $H_1 \subseteq H_2$._

Proof. Let $a_1, a_2, \ldots, a_q$ be a base for $\underset{\sim}{h}_1$ over K and let $T_1, T_2, \ldots, T_q$ be indeterminates. Define L and ω as in Theorem 7. Then $\omega \in H_1^{(L)}$ and is a generic point of H_1. Furthermore, by Theorem 6 we also have $\omega \in H_2^{(L)}$. Now every point of H_1 is a specialization of ω and therefore, by Chapter 2 Theorem 34 Cor. 2 applied to U(A), every such point belongs to H_2. Accordingly $H_1 \subseteq H_2$.

7.4 Applications to affine groups

We begin by generalizing Lemma 4.

Theorem 8. Let G be an affine group defined over a field K of characteristic zero, and let H_1 and H_2 be closed connected subgroups of G with $\underset{\sim}{h}_1$ and $\underset{\sim}{h}_2$ as their Lie algebras. (The Lie algebras $\underset{\sim}{h}_1$ and $\underset{\sim}{h}_2$ are to be regarded as subalgebras of $\underset{\sim}{g}$.) Then the following statements are equivalent:

(a) $H_1 \subseteq H_2$;

(b) $\underset{\sim}{h}_1 \subseteq \underset{\sim}{h}_2$.

Hence $H_1 = H_2$ if and only if $\underset{\sim}{h}_1 = \underset{\sim}{h}_2$.

Proof. It is clear that (a) implies (b). From here on we assume that $\underset{\sim}{h}_1 \subseteq \underset{\sim}{h}_2$.

Let Ω be the algebraic closure of K. By Theorem 27 of Chapter 5, we can regard $G^{(\Omega)}$ as a closed subgroup of $U(A)$, where A is an Ω-algebra of the kind discussed in the last section. Thus $H_i^{(\Omega)} \subseteq G^{(\Omega)} \subseteq U(A)$ for $i = 1, 2$. Let $\underset{\sim}{h}_i^*$ and $\underset{\sim}{g}^*$ denote the Lie algebras of $H_i^{(\Omega)}$ and $G^{(\Omega)}$ respectively. Then

$$\underset{\sim}{h}_i \subseteq \underset{\sim}{h}_i^* \subseteq \underset{\sim}{g}^* \subseteq A$$

and we know, from the discussion following Chapter 6 Theorem 20, that $\underset{\sim}{h}_i^*$ is the Ω-subspace of A spanned by $\underset{\sim}{h}_i$. Since $\underset{\sim}{h}_1 \subseteq \underset{\sim}{h}_2$, it follows that $\underset{\sim}{h}_1^* \subseteq \underset{\sim}{h}_2^*$ and therefore $H_1^{(\Omega)} \subseteq H_2^{(\Omega)}$ by Lemma 4. But, by Chapter 2 Theorem 30 Cor. 2, $H_i = G \cap H_i^{(\Omega)}$ and therefore $H_1 \subseteq H_2$ as required.

At this point we interrupt our discussion of affine groups in order to introduce the idea of a module over a Lie algebra.

To this end let Λ be a Lie algebra over K (K is an arbitrary field) and let V be a vector space over K (the dimension of V need not be finite). Then $\mathrm{End}_K(V)$ has a natural structure as a Lie algebra. Any homomorphism

$$\chi : \Lambda \to \mathrm{End}_K(V) \qquad (7.4.1)$$

of Lie algebras is called a representation of Λ by means of endomorphisms of V. For example, the adjoint representation[†] is of this kind.

Suppose that we have such a representation of Λ and put $\lambda v = (\chi(\lambda))(v)$ for all $\lambda \in \Lambda$ and $v \in V$. Then, with a self-explanatory notation,

$$\left.\begin{aligned} &\lambda(v_1 + v_2) = \lambda v_1 + \lambda v_2, \\ &\lambda(kv) = k(\lambda v) = (k\lambda)v \quad \text{for } k \in K, \\ &(\lambda_1 + \lambda_2)v = \lambda_1 v + \lambda_2 v, \\ &[\lambda_1, \lambda_2]v = \lambda_1(\lambda_2 v) - \lambda_2(\lambda_1 v). \end{aligned}\right\} \qquad (7.4.2)$$

It will be convenient to describe the properties set out in (7.4.2) by saying that the K-space V is a (left) module with respect to the Lie algebra Λ. Evidently the notion of a Λ-module is equivalent to that of a representation of Λ by endomorphisms of a K-space.

Let us introduce some further terminology. Suppose that Λ is a Lie algebra over K and that V is a Λ-module. By a Λ-submodule of V we mean a K-subspace W which satisfies $\lambda w \in W$ for all $\lambda \in \Lambda$ and $w \in W$. Again if $V \neq 0$ and it has no Λ-submodules apart from 0 and V itself, then we say that V is a simple Λ-module. Particularly important for us is the

Definition. The Λ-module V is said to be 'completely reducible' if for every Λ-submodule W, of V, there exists a second Λ-submodule W' such that $V = W \oplus W'$ this being a direct sum of K-spaces.

Now suppose that G is an affine group over K and that V is a finite-dimensional rational G-module.[‡] Then, by the corollary to Lemma 3 in Chapter 5, there results a rational representation

$$\rho : G \to GL(V) \qquad (7.4.3)$$

of G. This in turn gives rise to a homomorphism

---

† See section (6.2).

‡ For the remainder of Chapter 7, all G-modules are to be understood to be left G-modules.

$$d\rho : \mathfrak{g} \to \mathrm{End}_K(V) \tag{7.4.4}$$

of Lie algebras. Thus the finite-dimensional rational G-module V has a natural structure as a $\mathfrak{g}$-module.

Now suppose that K is an infinite field, let W be a K-subspace of V and consider GL(W, W), where the notation is the same as that employed in Example 7 of section (5.3). By Chapter 5 Theorem 18, GL(W, W) is a closed connected subgroup of GL(V). Also, as is easily verified from the definition,

$$GL(W, W) = \{\sigma \mid \sigma \in GL(V) \text{ and } \sigma W \subseteq W\}.$$

Consequently *W is a G-submodule of V if and only if* $\rho(G) \subseteq GL(W, W)$. Assume that this is the case. Then $d\rho(\mathfrak{g}) \subseteq \mathfrak{gl}(W, W)$ and therefore, by Chapter 6 Theorem 28, $\lambda w \in W$ whenever $\lambda \in \mathfrak{g}$ and $w \in W$. Accordingly *if W is a G-submodule of V, then it is also a $\mathfrak{g}$-submodule of V.*

As the next theorem shows, provided that K has characteristic zero and G is connected, the converse holds.

Theorem 9. *Let G be a connected affine group defined over a field K of characteristic zero and let V be a finite-dimensional rational G-module. Then the following statements are equivalent:*

(i) *W is a G-submodule of V,*

(ii) *W is a $\mathfrak{g}$-submodule of V.*

Proof. We shall assume (ii) and deduce (i). Our previous remarks show that this will be sufficient.

Let $\rho : G \to GL(V)$ be the rational representation to which V gives rise and let H be the closure of $\rho(G)$ in GL(V). Then H is a closed connected subgroup of GL(V) and there is induced an almost surjective K-homomorphism $G \to H$ of affine groups. Since the characteristic of K is zero, Lemma 2 of Chapter 6 shows that the induced Lie homomorphism $\mathfrak{g} \to \mathfrak{h}$ is a surjection and therefore $d\rho(\mathfrak{g}) = \mathfrak{h}$. But we are assuming (ii) to be true and this implies that $d\rho(\mathfrak{g}) \subseteq \mathfrak{gl}(W, W)$. Accordingly $\mathfrak{h} \subseteq \mathfrak{gl}(W, W)$ and so, by Theorem 8,

$$\rho(G) \subseteq H \subseteq GL(W, W).$$

Finally it follows from this that W is a G-submodule of V as required.

Theorem 10. *Let G be a connected affine group defined over a field of characteristic zero and let* $\mathfrak{g}$ *be its Lie algebra. Further let V be a finite-dimensional rational G-module. Then V is a completely reducible respectively simple G-module if and only if it is a completely reducible respectively simple* $\mathfrak{g}$*-module.*

This is an immediate consequence of Theorem 9.

Theorem 11. *Let G be a connected affine group defined over a field of characteristic zero. Assume that every finite-dimensional* $\mathfrak{g}$*-module is completely reducible. Then G is linearly reductive.*

Proof. Let V be a finite-dimensional rational G-module. By Theorem 10, V is a completely reducible G-module. That G is linearly reductive follows from Theorem 32 of Chapter 5.

These last results are powerful tools which can be used to show that certain special affine groups are linearly reductive. This is because they enable the classical theory of Lie algebras to be employed.

References

1. Borel, A. Linear algebraic groups, W. A. Benjamin Inc. (1969).
2. Chevalley, C. Théorie des groupes de Lie, Tome 2, Groupes algébriques, Hermann, Paris (1951).
3. Chevalley, C. Fundamental concepts of algebra, Academic Press (1956).
4. Fogarty, J. Invariant theory, W. A. Benjamin Inc. (1969).
5. Goldman, O. Hilbert rings and the Hilbert Nullstellensatz, Math. Zeit. 54 (1951), 136-40.
6. Hochster, M. and Eagon, J. A. Cohen Macaulay rings, invariant theory, and the generic perfection of determinantal loci, Amer. Journ. Math. 93 (1971), 1020-58.
7. Krull, W. Jacobsonches Radikal und Hilbertscher Nullstellensatz, Proc. International Congress of Mathematicians Vol. 2, Cambridge Mass. (1950), 54-64.
8. Nagata, M. Local rings, Interscience Tracts in Pure and Applied Mathematics, No. 13, Interscience Publishers (1962).
9. Northcott, D. G. Lessons on rings, modules and multiplicities, Cambridge Univ. Press (1968).
10. Van der Waerden, B. L. Moderne Algebra, Zweiter Teil, Springer (1931).
11. Weil, A. Foundations of algebraic geometry, Amer. Math. Soc. Colloquium Publications, Vol. 29 (1946).
12. Zariski, O. The concept of a simple point on an abstract algebraic variety, Trans. Amer. Math. Soc. 62 (1974), 1-52.
13. Zariski, O. and Samuel, P. Commutative Algebra, Vols. 1, 2, Univ. Series in Higher Mathematics, van Nostrand (1960).

Index

additive group of the ground field 150

adjoint representation of a Lie algebra 227

adjoint representation of an affine group 256

affine group 134

affine K-algebra 27

affine n-space 30

affine set 24

affine topology 24

almost surjective K-morphism 84

associated ideal of a set of points 21

associative K-algebra 3

Basis Theorem 64

completely reducible module 176

completely reducible module relative to a Lie algebra 280

connected component of the identity 143

coordinate function 30

coordinate ring of an affine set 24

derivation 95

derived function algebra 18

differential of a rational function 126

dimension of an affine set 75

dual base 29

dual space 29

finitely generated algebra 5

finite sets topology 23

formal power series 263

function algebra 17

generalized point 55

general K-algebra 222

general liner group 33

general point 56

Hilbert ring 25

homomorphism of affine groups 159

homomorphism of K-algebras 4

invariant derivation 227

irreducible affine set 64

irreducible components of a space 62

irreducible space 60

isomorphism of affine sets 35

Jacobi's identity 224

Jacobson ring 25

left translation 136

Lie algebra 224

Lie algebra of an affine group 228

Lie product 225

linear form on a vector space 29

linearly disjoint fields 82

linearly reductive affine group 176
local derivation 115
local homomorphism 74
local ring of an affine set at a point 71
locus of a set of functions 19

minimal prime ideals of an ideal 67
module with respect to a Lie algebra 280
morphic action of an affine group on an affine set 162
morphism of affine sets 34
multiple point 119
multiplicative group of the ground field 149

natural prolongation of a coordinate function 48
natural prolongation of a rational function 83
Noetherian ring 64
Noetherian space 63
non-associative algebra 222
non-trivial K-algebra 4

orbit 164
orthogonal group 159

principal locus 20
products of affine sets 40
products of function algebras 38

quotient for the action of an affine group 196

rational character 181
rational function on an irreducible affine set 75
rational G-module (finite-dimensional case) 167
rational G-module (general case) 174
rationally reduced K-algebra 15
rational maximal ideal 13
rational representation 169
regular extension of a field 82
regular left action 163
regular right action 163
representation of a Lie algebra 279
Reynold's operator 191
right translation 136

semi-invariant 183
separable extension of a field 108
separable K-homomorphism 237
separating transcendence base 108
simple G-module 175
simple point 119
specialization 56
special linear group 150
strict quotient for the action of an affine group 200
structural homomorphism of an algebra 4
symplectic group 159

tangent space at a point 115
tangent vector 117
tensor products of algebras 8
tensor products of vector spaces 6
torus 149

unitary K-algebra 3

weight of a semi-invariant 183